NVIVO 10 ESSENTIALS

Your Guide to the World's Most Powerful Data Analysis Software

By Bengt M. Edhlund & Allan G. McDougall

FORM & KUNSKAP AB • P.O. BOX 4 • SE-645 06 STALLARHOLMEN • SWEDEN • +46 152 201 80
SALES@FORMKUNSKAP.COM • WWW.FORMKUNSKAP.COM

NVIVO
NVivo is a registered trademark of QSR International Pte Ltd.

WORD, EXCEL, ONENOTE
Word, Excel, and OneNote are registered trademarks of Microsoft Corporation.

ENDNOTE
EndNote is a registered trademark of Thomson Reuters.

EVERNOTE
Evernote is a registered trademark of Evernote Corporation.

LIMITED LIABILITY
This document describes the functions and features valid for NVivo 10 in combination with Word 2003, 2007 or 2010. Form & Kunskap is unable to predict or anticipate changes in service performance or functionality of later versions or changes of respective software that might influence or change the described features.

COPYRIGHT FORM & KUNSKAP AB 2012
The content of this document is subject to regulation according to the Swedish law on copyright, 1960:729. The said contents must not be reproduced or distributed in any form without a written consent from Form & Kunskap Stockholm Aktiebolag. The restrictions are valid for the entire document or parts thereof and include electronic storage, display on VDU or a similar device and audio or video recording.

ISBN 978-1-300-04132-0

FOREWARD

NVivo 10 Essentials is our comprehensive guide to the world's most popular qualitative data analysis software. This book has two authors. Bengt Edhlund is a software trainer with decades of experience teaching and writing about research software. Allan McDougall has worked with NVivo 8, NVivo 9, and NVivo 10 as a qualitative health researcher and doctoral student. We have co-authored this book to provide instruction to NVivo users of all skill levels and experience with both qualitative data analysis and qualitative research methods will benefit from this book. We break down the functional components of this intricate software. We also strive to provide practical, anecdotal advice for *using* NVivo 10 for every stage of your research project. Further, we strive to provide advice on using NVivo 10 in a collaborative environment. A topic on which we are aware many of our colleagues are interested. Please enjoy our book and feel free to contact us at any time at: info@formkunskap.com

Form & Kunskap AB, founded by Bengt in 1993, is a training company focused on software solutions for academic researchers: We believe any single product cannot be the sole solution for researchers and research teams. The best outcomes will always be the result of the combination of several products.

We believe it is important to always select market leading software products that follow recognized industry standards. We believe in teaching vertically, cutting across the planes of single software solutions. We provide our clients with well written literature followed by professional support. Many years of teaching and support has helped us understand how to teach academics to use tools that are both complex and productive.

TABLE OF CONTENTS

1. INTRODUCING NVIVO 10 11
   Welcome to NVivo 10 Essentials 11
   What is NVivo 10? 12
   Exploring this Book 15
   Graphic Conventions 16
   Aspects on Installation of NVivo 10 16
   System Requirements – Minimum 17
   System Requirements – Recommended 17
   Requirements for Macs 17
   What's New in NVivo 10? 18
2. THE NVIVO 10 INTERFACE 19
   Area 1 – The Navigation Window 20
   Area 2 – The Virtual Explorer 21
   Creating a New Folder 21
   Deleting a Folder 22
   Area 3 – The List View 22
   Area 4 – The Detail View 26
   Copying, Cutting, and Pasting 28
   Creating New Sets 29
   Undo 31
   The Ribbon 32
   Application Options 36
   Alternate Screen Layouts 47
3. BEGINNING YOUR PROJECT 49
   Creating a New Project 49
   Sources & Project Size 50
   Project Properties 51
   Merging Projects 60
   Exporting Project Data 60
   Exporting Project Items 61
   Save and Security Backup 62
4. HANDLING TEXT SOURCES 63
   Documents 63
   External Items 68
5. EDITING TEXT IN NVIVO 71
   Formatting Text 71
   Aligning Paragraphs 72
   Finding, Replacing and Navigating Text 73
   Searching and Replacing 74
   Spell Checking 74
   Selecting Text 76
   Go to a Certain Location 77

	Creating a Table	77
	Inserting Page Breaks, Images, Dates, and Symbols	78
	Zooming	78
	Print Previewing	79
	Printing a Document	80
	Printing with Coding Stripes	81
	Page Setup	82
	Limitations in Editing Documents in NVivo	83
6.	HANDLING PDF SOURCES	85
	Importing PDF Files	85
	Opening a PDF Item	87
	Selection Tools for PDF Items	87
	Exporting a PDF Item	88
7.	HANDLING AUDIO- AND VIDEO-SOURCES	91
	Importing Media Files	92
	Creating a New Media Item	94
	Opening a Media Item	95
	Play Modes	97
	Coding a Media Item	103
	Working with the Timeline	104
	Linking from a Media Item	105
	Exporting a Media Item	105
8.	HANDLING PICTURE SOURCES	107
	Importing Picture Files	107
	Opening a Picture Item	110
	Selecting a Region and Creating a Picture Log	111
	Editing Pictures	112
	Coding a Picture Item	112
	Linking from a Picture Item	114
	Exporting a Picture Item	115
9.	MEMOS, LINKS, AND ANNOTATIONS	117
	Exploring Links in the List View	117
	Memos	117
	See Also Links	122
	Annotations	125
	Hyperlinks	126
10.	INTRODUCING NODES	129
	Case Nodes and Theme Nodes	129
	Creating a Node	131
	Building Hierarchical Nodes	132
	Merging Nodes	133
	Relationships	134
11.	CLASSIFICATIONS	139
	Node Classifications	140
	Source Classifications	140

- Creating a Classification 140
- Customizing a Classification 142
- Working with the Classification Sheet 144
- Exporting Classification Sheets 147
- Importing Classification Sheets 148

12. **CODING** 153
    - The Quick Coding Bar 153
    - Drag-and-Drop Coding 155
    - Menus, Right-Click, or Keyboard Commands 155
    - Autocoding 158
    - Range Coding 160
    - In Vivo Coding 161
    - Coding by Queries 161
    - Visualizing your Coding 161
    - Viewing a Node that Codes a PDF Source Item 174

13. **QUERIES** 177
    - Word Frequency Queries 178
    - Text Search Queries 181
    - Coding Queries 187
    - Compound Queries 193
    - Matrix Coding Queries 196
    - Group Queries 206

14. **COMMON QUERY FEATURES** 209
    - The Filter Function 209
    - Saving a Query 210
    - Saving a Result 211
    - About the Results Folder 212
    - Editing a Query 213
    - The Operators 215

15. **HANDLING BIBLIOGRAPHIC DATA** 219
    - Importing Bibliographic Data 221
    - The PDF Source Item 223
    - The Linked Memo 225

16. **ABOUT THE FRAMEWORK METHOD** 227
    - Introducing the Framework Matrix 227
    - Creating a Framework Matrix 229
    - Populating Cell Content 231
    - Working with Framework Matrices 236
    - Organizing Framework Matrices 239

17. **ABOUT QUESTIONNAIRES AND DATASETS** 241
    - Importing Datasets 241
    - Exporting Datasets 248
    - Coding Datasets 248
    - Autocoding Datasets 249
    - Classifying Datasets 253

18. INTERNET AND SOCIAL MEDIA 257
    Introducing NCapture 257
    Exporting websites with NCapture 257
    Importing Websites with NCapture 259
    Social Media Data and NCapture 260
    Exporting social media data with NCapture 262
    Importing Social Media Data with NCapture 262
    Working with Social Media Datasets 264
    Analyzing Social Media Datasets 266
    Autocoding a Dataset from Social Media 266
    Installing NCapture 270
    Check your Version of NCapture 270
19. USING EVERNOTE WITH NVIVO 271
    Evernote for Data Collection 271
    Exporting Notes from Evernote 271
    Importing Evernote Notes into NVivo 272
    Evernote Note Formats in NVivo 273
    Autocoding your Evernote Tags 273
20. USING ONENOTE WITH NVIVO 275
    Exporting Notes from OneNote 275
    Importing OneNote Notes into NVivo 276
    OneNote Note Formats in NVivo 277
    Installing NVivo Addin for OneNote 277
    Check whether NVivo Addin for OneNote is Installed 277
21. FINDING AND SORTING PROJECT ITEMS 279
    Find 279
    Advanced Find 280
    Sorting Items 285
22. COLLABORATING WITH NVIVO 10 287
    Current User 287
    Viewing Coding by Users 289
    Viewing Coding Stripes 289
    Coding Comparison Query 290
    Models and Reports 293
    Tips for Teamwork 294
    A Note on Cloud-computing 295
    A Note on NVivo Server 295
23. MODELS 297
    Creating a New Model 297
    Display Color Codes 300
    Creating a Static Model 301
    Creating Model Groups 301
    Adding More Graphical Shapes 302
    Creating Connectors between Graphical Items 303
    Deleting Graphical Items 304

Converting Graphical Shapes	304
Editing a Graphical Item	304
Exporting your Model	306
24. MORE ON VISUALIZING YOUR DATA	**307**
Cluster Analysis	307
Tree Maps	311
Graphs	313
25. REPORTS AND EXTRACTS	**315**
Understanding Views and Fields	315
Report and Extract Templates	316
Reports	317
Extracts	326
26. HELP FUNCTIONS IN NVIVO	**331**
Help Documents	331
Tutorials	332
Support and Technical Issues	332
Software Versions and Service Packs	333
27. GLOSSARY	**335**
APPENDIX A – THE NVIVO SCREEN	**341**
APPENDIX B – KEYBOARD COMMANDS	**343**
INDEX	**347**

1. INTRODUCING NVIVO 10
Welcome to NVivo 10 Essentials

Welcome to NVivo 10 Essentials, your guide to the world's most powerful qualitative data analysis software. The purpose of this book is to provide a comprehensive guide to every feature of NVivo 10. For beginners, you will find explanations of key concepts and recommendations for starting your first NVivo project, importing your data, analyzing your data, and sharing your findings with collaborators. Some people find it hard to wrap their heads around NVivo, and you might be one of those people. Perhaps you have been playing around with the software already and you don't really 'get it'. This book offers that one simple description of what to do with NVivo and how you can make it work for you. Learning NVivo is like learning Microsoft Excel, it is a complex software tool that can be applied to approach many types of problems. There is no one right way to use NVivo, but we the authors have enough collective experience to offer some best practices that will help you along your journey.

For advanced users, you will find a comprehensive introduction to NVivo 10's new features, including the ability to work with social media data. You may wish to skip to page 18 for our summary, What's New in NVivo 10?

Whatever your skill level using NVivo 10 is, this book has been written by two authors who combine theory and practice to offer readers a guide towards both technical capability and practical application.

Bengt M. Edhlund

Bengt Edhlund is the author of several books, including *NVivo 9 Essentials*. He is Scandinavia's leading research software trainer. As a former telecommunications engineer, Bengt has published 7 books on academic informatics tools such as NVivo, EndNote, PubMed, and Excel. All of Bengt's books are available in English and Swedish. He has trained researchers from every corner of the globe, including Canada, Sweden, Norway, China, Egypt, Uganda, and Vietnam. A trainer who takes pride in his students' success, Bengt provides all of his clients with customized NVivo support solutions via Skype or email.

Allan G. McDougall

A former student of Bengt's, Allan is a qualitative researcher with extensive NVivo experience. He has used NVivo in a collaborative academic environment on a number of diverse projects related to his area of qualitative health research. While Bengt knows every facet of NVivo's various functions, Allan is an NVivo user who provides practical tips from his personal experiences throughout this book.

What is NVivo 10?

NVivo 10 allows users to work with a variety of data, such as documents, images, audio, video, questionnaires and web / social media content.

Whether you apply grounded theory, phenomenology, ethnography, discourse analysis, attitude surveys, organizational studies or mixed-methods you will soon be required to bring order and structure to your data. We have worked with researchers tasked with analyzing hundreds of interviews and focus group discussions, or researchers tasked with analyzing thousands of short-answer questionnaires. Although at one time researchers were able to analyze data like this using paper-based systems, today it is faster, easier and more efficient to use NVivo to verify theories and collaborate.

Key Components of an NVivo 10 Project

The purpose of this section is familiarizing you with some of the terminology you will need to understand to work with NVivo 10. **Sources**, **Nodes**, and **coding** are the building blocks of your project. You will learn in-depth explanations of these concepts both in the logical journey through the chapters of this book, but also in our glossary that appears at the end of the book. Below is a simplified diagram of an NVivo project's key components:

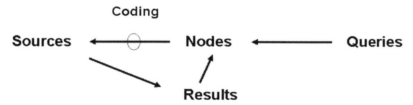

Within an NVivo **Project**, NVivo helps you conceptualize data organization in terms of **Project Items** and **Folders**. Items and folders in NVivo are 'virtual' in relation to a Windows-based or Mac-based environment. In an NVivo Project folders are similar to Windows folders, but NVivo has set up some rules for how they are handled (for example only certain types of folders are allowed to be organized hierarchically). Project Items correspond to files in Windows environment and are handled in most cases like them. They may be edited, copied, cut, pasted, deleted, moved, etc., but they are all embedded in the project file.

Sources are your data; they can be documents, audio, video and image files, memos and other external items. When Sources are imported or linked to NVivo they are mirrored as NVivo Project Items. Items like documents can also be created directly in NVivo's own word processor.

Memos are also items in a project. A memo is linked to a particular document or Node. They can also be imported, but often they are created in NVivo. **Links** of various kinds can be created between internal items and even outside of NVivo to external sources.

A unique feature of NVivo is that certain types of items can be arranged hierarchically, namely Nodes. **Nodes** are terms and concepts that are created during the process to designate properties, phenomena, or keywords that characterize sources or individual parts of the sources. Nodes can also represent tangible components of your project, such as study participants or geographic locations. We would call the former type of Node a **Theme Node** and the later type of Node a **Case Node**. Nodes are of different types: parent Nodes, child Nodes, relationships, and matrices.

Coding is the activity which means that words, sentences, whole paragraphs, other graphic elements or the whole object is associated with some Nodes. Coding can take place only by items that are included in the project. An external source cannot be encoded but the text is in the external object can be encoded.

Queries can recognize which parts of the project's sources that contain the desired information. A simple query is simply opening a Node. Exploring more complex issues can involve combinations of Nodes that are linked with Boolean operators. Queries can be saved to be reused as the projects develop. Results of queries can also be saved or be used to generate new Nodes. Queries may also be in matrix form so that the rows of some Nodes and columns of other Nodes create a table where each cell is the result of two Nodes and a particular operator.

An overall picture of how a project can be developed is:

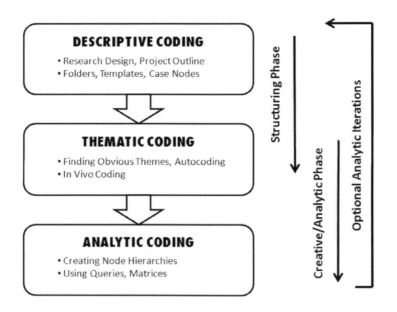

NVivo helps to organize data so that analysis and conclusions will be safer and easier. The ultimate goal may be described as follows:

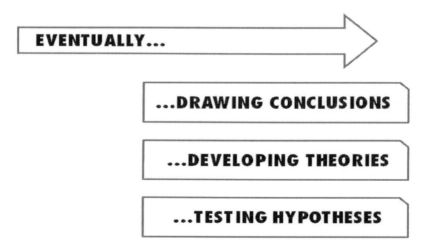

- ♦ -

Exploring this Book

This book is structured so that we start with the system requirements for software in this chapter. We describe in Chapter 2 how NVivo interface is designed and the basic settings you can do to optimize your NVivo software itself. Chapter 3 explains how to create, save and backup your NVivo Project file.

Chapters 4 - 8 cover how to import, create and edit text, audio, video, and picture data. Chapter 9 outlines how to create memos and links. Chapters 10 and 11 explore the lifeblood of qualitative data analysis: Nodes, classifications and coding. Chapters 13 and 14 discuss the act of creating queries, saving them and creating Nodes from query results.

Chapter 15 deals with NVivo's powerful functionality to support literature reviews and bibliographic data, with Chapter 16 describing the Framework Method of conveniently viewing and summarizing your data in Framework Matrices. Chapter 17 moves into managing surveys and questionnaires (now called Datasets in NVivo).

Chapter 18 covers NVivo 10's newest and most exciting feature: capturing data from the web and social media. Chapter 19 and 20 describes an equally exciting new feature, NVivo's new functionality to work with Evernote and OneNote – if you don't know Evernote we provide a brief introduction.

Chapter 21 describes finding and sorting various items within an NVivo project. Chapter 22 deals with important aspects of using NVivo on a research team. In a related chapter, Chapters 23 and 24 describe how to graphically illustrate a project using Models and Charts. And Chapter 25 describes the reporting methods.

Finally, Chapter 26 reviews the help functions available in NVivo 10, and Chapter 27 contains a glossary that occurs in connection with NVivo.

Graphic Conventions

In this book we have applied some simple graphic conventions with the intention of improving readability:

Example	Comment
Go to **Model** | **Shapes** | **Change Shape**	Ribbon menu **Model** och Menu group **Shapes** and Menu option **Change Shape**
Go to **File** → **Options**	Main menu and options with **Bold**
Right-click and select **New Query** → **Compound...**	Right-click with the mouse and select menu and sub-menu with **Bold**
Select the **Layout** tab	Optional tabs with **Bold**
Select *Advanced Find* from **Options** drop-down list	Variable with **Bold**, the value with *Italic;* Heading with **Bold**, options with *Italic*
Confirm with **[OK]**	Graphical buttons within brackets
Use the **[Del]** key to delete	Key is written within brackets
Type `Bibliography` in the textbox	`Courier` for text to be typed
`..[1-3]` is shown in the textbox	`Courier` for shown text
.. key command **[Ctrl]** + **[Shift]** + **[N]**	Hold the first (and second) key while touching the last

Aspects on Installation of NVivo 10

Your present Windows version decides if you should install 32-bit or 64-bit version of NVivo 10. The 32-bit version can be installed on all the below mentioned Windows versions. The 64-bit version can only be installed when you have a 64-bit version of Windows 7. If possible we recommend to use 64-bit version of NVivo 10 which is faster to download and install and also has a somewhat better performance.

Installation is made in two steps: The 'installation' which requires a License key and then the 'activation' which requires a communication with QSR preferably over the internet. The activation registers the user data in a customer database at QSR and is a license control function.

Should you need to change computer at a later stage then you need to deactivate your license **before** you uninstall NVivo in the old machine. Then you can reinstall NVivo in the new computer with your license key followed by a new activation procedure.

System Requirements – Minimum

- 1,2 GHz Pentium III-compatible processor or faster (32-bit); 1,4 GHz Pentium 4-processor (64-bit)
- 1 GB RAM or more
- 1024 x 768 screen resolution or higher
- Microsoft Windows XP SP2 or later
- Approximately 2 GB of available hard-disk space

System Requirements – Recommended

- 2 GHz Pentium 4-compatible processor or faster
- 2 GB RAM or more
- 1280 x 1024 screen resolution or higher
- Microsoft Windows XP SP 2 or later;
  Microsoft Windows Vista SP 1 or later;
  Microsoft Windows 7
  Microsoft Windows 8
- Approximately 2 GB of available hard-disk space
- Internet Explorer 7 or later (for NCapture)
- Internet Connection

We recommend that your machine complies with those recommendations required for larger projects even if you are working with smaller projects.

Requirements for Macs

A Mac version of NVivo 10 is not currently on the market, however NVivo 10 will operate on a Macintosh (Mac) computer. First, the user must set up some method of accessing Windows on their Mac. This can be done using dual-boot software (where a max computer has both Windows and a Mac operating system installed) or using virtualization software (where the Mac 'tricks' NVivo 10 into believing it is operating on a Windows computer). Historically, NVivo 10 Mac users have found three tools most useful: Boot Camp, Parallels and VMware Fusion.

Boot Camp

Boot Camp is a dual-boot software utility that has been included in Mac operating systems since 2007. NVivo 10 Mac users utilizing Boot Camp must ensure their system requirements specified for NVivo 10 on Windows should be met. Again, Boot Camp allows Mac users to 'boot up' their Mac computer to run either a Mac operating system or a Windows operating system. Boot Camp is not popular with all NVivo 10 Mac users because it requires users switch between operating systems to use NVivo 10. Conversely, Boot Camp is an included Mac feature and there is free to run.

Parallels & VMware

Parallels and VMware are virtualization software that ostensibly 'trick' NVivo 10 into believing it is operating on a Windows-based computer. The advantage for Mac users here is that, unlike Boot Camp, they are not required to switch between a Mac and Windows operating system on their computer. However, Mac users should be aware that when using Parallels or VMware Fusion (or similar products), they may be required to have a Mac computer with higher system specifications than those detailed above. These emulation programs require increased memory and processor functions because they are simultaneously running a Mac operating system and a 'virtual' Windows system as well. Feel free to contact the authors if you have any questions about whether or not your computer's resources pose and challenges for NVivo 10 functionality.

What's New in NVivo 10?

You'll learn all about what's new in NVivo 10 throughout this book. But we thought it would be important for our more advanced readers to have a go-to guide to find out the ins and outs of NVivo's new features:

- Capturing webpages as data (Chapter 18)
- Capturing social media data from Facebook, Twitter and LinkedIn (Chapter 18)
- Full integration of Evernote notes and OneNote pages (Chapter 19 & 20)
- Spellchecking your text Sources (page 74)
- Increased audio and video file formats (Chapter 7)
- Print with coding stripes on same page (page 81)
- Reports and Extracts including Contents (Chapter 25)

2. THE NVIVO 10 INTERFACE

This chapter is about the architecture of the NVivo screen which resembles the interface of Microsoft Outlook.

Appendix A, The NVivo Screen (see page 341) shows an overview of the main NVivo window. We will use Area 1, Area 2 etcetera to represent the various sectors of the NVivo window. A work session usually starts in Area 1 with the selection of a Navigation Button corresponding to a group of folders. In Area 2, you select the folder relevant to your analysis, which leads to Area 3 where you can select a certain Project Item. Is Area 4 where you can study that item's content.

Project work is done through the Ribbon menus, keyboard commands or or via the menu options brought up by right-clicking your mouse. For more information about these commands Appendix B (see page 343) is a summary of NVivo's keyboard commands.

> **Tip:** We suggest simply opening your NVivo 10 software and experimenting with the interface. Challenging yourself to play around in NVivo is a great way to learn!

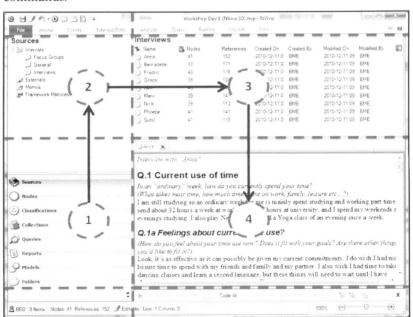

You will find the Status Bar below the four areas comprising the main NVivo screen. The Status Bar displays information (depending on the cursor position) on number of items in the current folder, number of Nodes and references of the current item, and row number and column number of the current cursor position.

Area 1 – The Navigation Window

Area 1, the Navigation Window contains 8 navigation buttons. In the bottom right corner the >> symbol reveals options on how to hide certain buttons and how to change their display order. Each button will display certain preselected folders in Area 2 and the button [**Folders**] displays all folders.

In case a certain button that you need should be currently hidden you can always go to **Home | Workspace | Go** and from there to any navigation option. This menu also shows the corresponding keyboard commands.

Area 2 – The Virtual Explorer

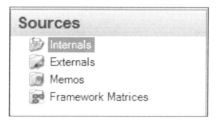

Each Navigation button contains a number of folders where relevant Project Items are stored. The folders associated with each Navigation Button are displayed in the Virtual Explorer, Area 2.

Virtual file paths are called Hierarchical Names. Only folders and Nodes have hierarchical names in NVivo 10. In the NVivo environment, hierarchical names are written with a double backslash between folders and a single backslash between a Node and its child Node.

For example:

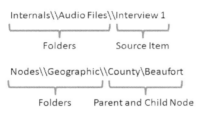

> **Did you know?** NVivo folders are called *virtual folders* as opposed to folders in a Windows environment. NVivo folders are *virtual* because they exist only in the NVivo project file. In most cases, *Virtual folders* perform like any Windows folder - you can create sub-folders, drag and drop project items into allowable folders, copy and paste folders. Certain folders are predefined in the NVivo project template and cannot be changed or deleted whereas other folders can be created by the user.

Creating a New Folder

NVivo contains a core set of template folders that cannot be deleted or moved. Users can create new subfolders under some of these template folders: Internals, Externals, Memos, Framework Matrices, Nodes, Relationships, Node Matrices, Source Classifications, Node Classifications, Relationship Types, Sets, Search Folders, Memo Links, See Also Links, Annotations, Queries, Results, Reports, Extracts, and Models:

1. Select one of the navigation buttons in Area 1 and then select the folder in Area 2 under which a new subfolder will be created.
2. Go to **Create | Collections | Folder**
   or right-click and select **New Folder...**
   or [**Ctrl**] + [**Shift**] + [**N**].

For each new folder, the **New Folder** dialog box appears:

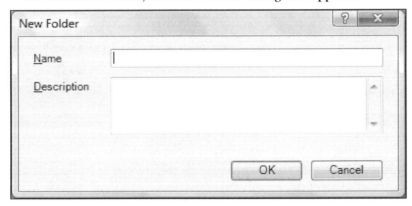

3   Type a name (compulsory) and a description (optional), then [**OK**].

Deleting a Folder

Deleting a folder also deletes its subfolders and all contents (all items in Area 3) therein.

1   Select the folder or folders in Area 2 that you want to delete.
2   Go to **Home | Editing | Delete**
    or right-click and select **Delete**
    or [**Del**].
3   Confirm with [**Yes**].

Area 3 – The List View

Area 3 appears similar to a list of files in Windows, but NVivo calls these Project Items within an Item List. All folders are Project Items. All Project Items in the Internals, Externals and Memos folders - and their subfolders - are *Source Items*.

During the course of a project, you may need to revise the item lists when items are created, deleted or moved. At times it may be necessary to refresh the item list:

1   Go to **Home | Workspace | Refresh**
    or [**F5**].

Item Properties

All items have certain characteristics that can be changed or updated through the item's properties menu:
1. Select the item in Area 3 that you want to change or update.
2. Go to **Home | Item | Properties**
   or **[Ctrl] + [Shift] + [P]**
   or right-click and select <**Item type**> **Properties**.

An item properties dialog box (in this case, **Audio Properties**) may look like this:

All information in this dialog box is editable and the text in the text boxes **Name** and **Description** is also searchable with the **Find** function, see Chapter 21, Finding and Sorting Project Items. Most experienced NVivo users know the project properties shortcut off by heart because this is fastest way to reach an item's Descriptions, a profoundly useful function in NVivo we'll spend more time on in Chapter 22, Collaborating with NVivo 10.

Setting Colors

Source items, Nodes, relationships, attribute values or users can be color marked individually. NVivo has seven pre-defined colors. Colors serve as visual cues for project researchers and as a result they can be used for a number of reasons. Most importantly, a color assigned to a Node will be visually represented in coding stripes (see page 168). The color marking is shown in the List View of Area 3 and can also be shown in Models and other visualizations (see page **Error! Bookmark not defined.**).
1. Select the item or items that you want to color mark.
2. Go to **Home | Item | Properties** → **Color** → <select>
   or right-click and select **Color** → <select>.

Classifying an Item

All Source Items (except Framework Matrices) or Nodes (except Relationships and Matrices) can be classified and thus associated with meta-data. We'll discuss this further in Chapter 11, Classifications, but for now this is a reminder that working with items takes place in the List View, Area 3.

1. Select the item or items that you want to classify.
2. Go to **Home | Item | Properties** → **Classification** → <select>
or right-click and select **Classification** → <select>.

Viewing Options

The List View default view is shown above in the first figure of this section. But there are three more options for viewing items in the List View: Small, Medium and Large Thumbnails.

1. Click on any empty space in Area 3.
2. Go to **View | List View | List View** → <select>.

The result of choosing *Large Thumbnails* may look like this:

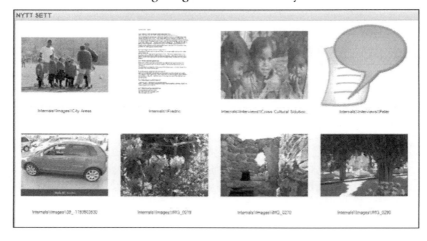

Sorting Options for a List
There are several ways to sort a list in Area 3.
1. Click on any empty space in Area 3.
2. Go to **Layout | Sort & Filter | Sort by** → <select>.

You can also perform a custom sort especially for nodes by moving items in the list.
1. Selec t one or more items in Area 3.
2. Go to **Layout | Rows & Columns | Move Up** ([**Ctrl**]+[**Shift**]+[**U**]).

alternatively
2. Go to **Layout | Rows & Columns | Move Down** ([**Ctrl**]+[**Shift**]+[**D**]).

Your custom sorting is saved so it can at any later stage be resumed by going to **Layout | Sort & Filter | Sort by** → **Custom**.

Customizing the Item List
You can customize the columns associated with Project Items. The **Customize Current View** dialog box allows you to remove unnecessary columns or add additional ones.
1. Click on any empty space in Area 3.
2. Go to **View | List View | List View** → **Customize...**

alternatively
2. Right-click and select **List View** → **Customize...**

> **Tip:** Some videos begin with a black frame, making the List View thumbnail a black square. But thumbnails of video items can display the specific frame that you want.
> 1. Move the playhead the frame you want to display.
> 2. Klick in the video fram.
> 3. Go to **Media | Selection | Assign Frame as Thumbnail**.
>
> The selected frame is displayed as a thumbnail in Area 3 and when using the Video tab for a Node that the Video Item is coded at.

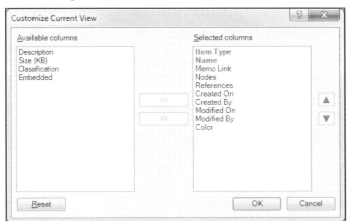

Printing the Item List
Printing the Item List can be a valuable contribution to discussions in project team meetings:
1. Go to **File → Print → Print List...**
   or right-click and select **Print → Print List...**
2. Select printer and printer settings, then **[OK]**.

Exporting the Item List
Exporting your item list as an Excel spreadsheet or a text document is also possible:
1. Go to **External Data | Export | Export → Export List...**
   or right-click and select **Export → Export List...**
2. Select file format, folder and name, then **[OK]**.

Deleting an Item
You can delete items from Area 3. When you delete a parent Node you also delete its child Nodes. Likewise, deleting a Classification also deletes its Attributes.
1. Click the appropriate navigation button in Area 1.
2. Select the appropriate folder in Area 2 or its subfolder.
3. Select the item or items in Area 3 that you want to delete.
4. Go to **Home | Editing | Delete**
   or right-click and select **Delete**
   or **[Del]**.
5. Confirm with **[Yes]**.

Area 4 – The Detail View

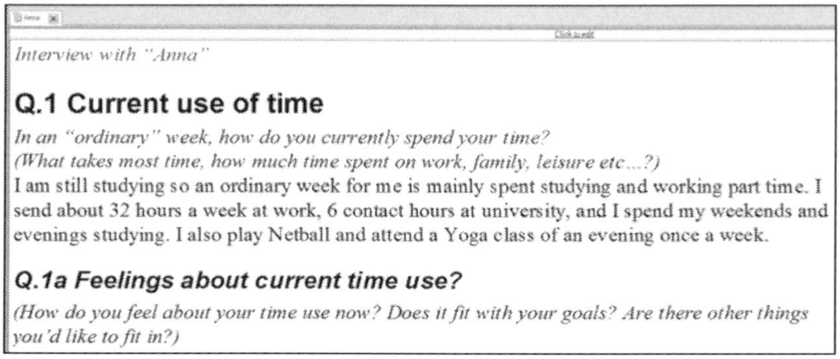

> *Interview with "Anna"*
>
> ### Q.1 Current use of time
> *In an "ordinary" week, how do you currently spend your time?*
> *(What takes most time, how much time spent on work, family, leisure etc....?)*
> I am still studying so an ordinary week for me is mainly spent studying and working part time. I send about 32 hours a week at work, 6 contact hours at university, and I spend my weekends and evenings studying. I also play Netball and attend a Yoga class of an evening once a week.
>
> ### Q.1a Feelings about current time use?
> *(How do you feel about your time use now? Does it fit with your goals? Are there other things you'd like to fit in?)*

The above image is an example of an open Project Item, which can include documents, audio clips, videos, pictures, memos, or Nodes.

> **Did you know?** Read Only mode doesn't mean that a project item is *locked* from making changes. You can still code and create links (but not hyperlinks) in a Read Only Source Item.

Each time a Source Item is opened it is Read Only. The document is made instantly editable by clicking on the Click to edit link at the top of an open item. Alternatively, you can go to **Home | Item | Edit** or **[Ctrl] + [E]** which is a toggling function.

Each item has its own tab when several items are opened at the same time. By default, Project Items are 'docked' inside Area 4. You can undock an open item as a standalone window:
 1 Go to **View | Window | Docked**.

Any undocked item can be docked again:
 1 Select the undocked item.
 2 Go to **View | Window | Docked**.

Undocking items can only take place during an open work session; when you reopen a project all undocked windows are closed. However, you can go to **File → Options** and in the **Application Options** dialog box, select the Display tab, under the Detail View Defaults section, beside Window, you can select *Floating* (see page 39) so that an item window is always opened in undocked mode.

> **Tip:** When you have several open project items, you can undock these items as seperate windows:
> 1 Go to **View | Workspace | Undock All**.
>
> Conversely, if you want to dock all of your open items, click outside any of the undocked items:
> 1 Go to **View | Workspace | Dock All**.
>
> While any docked window can be closed by clicking x on the right side of its tab, all windows can be closed simultaneously:
> 1 Go to **View | Workspace | Close All**.

Copying, Cutting, and Pasting

Standard conventions for copying, cutting, and pasting text and images prevail in NVivo. In addition, NVivo can also copy, cut, and paste complete Project Items like documents, memos, Nodes, etc. However, it is not possible to paste Nodes into folders meant for documents and vice versa (this would breach the software's folder template conventions). It is only possible to paste an item into the folder appropriate for that type of folder (e.g., paste a Node within the Nodes folder, a Query within the Queries folder, etc.). To cut and paste within NVivo:

1. Select an item (document, Node etc.)
2. Go to **Home | Clipboard | Cut**
or right-click and select **Cut**
or [**Ctrl**] + [**X**].

alternatively

2. Go to **Home | Clipboard | Copy**
or right-click and select **Copy**
or [**Ctrl**] + [**C**].
3. Select the appropriate folder or parent Node under which you want to place the item.
4. Go to **Home | Clipboard | Paste → Paste**
or right-click and select **Paste**
or [**Ctrl**] + [**V**].

Paste Special

The normal **Paste** command includes all those elements. But after copying or cutting of some items (excluding Nodes) you can decide which elements from the item that should be pasted:

1. Copy or cut the item or items that you want to paste into the new position.
2. Select the target folder.
3. Go to **Home | Clipboard | Paste → Paste Special...**

The **Paste Special Options** dialog box appears:

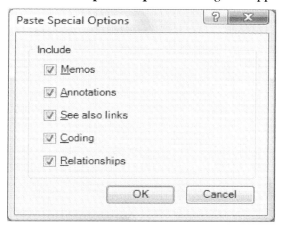

4   Select the item elements that you want to include. Additional context-based options may also appear: *Media content* and *Transcript* are valid for video and audio items and *Log entries* is valid for picture items.
5   Confirm with [**OK**].

Creating New Sets

Sets are defined as folders under the parent folder **Sets** that contains shortcuts to various Project Items or groups of Project Items. A set is considered a subset or collection of Project Items that allow you to access organized groups of items without moving or copying those items.

1   Go to [**Folders**] or [**Collections**] in Area 1.
2   Select the **Sets** folder in Area 2.
3   Go to **Create | Collections | Sets**
    or right-click and select **New Set...**
    or [**Ctrl**] + [**Shift**] + [**N**].

The **New Set** dialog box appears:

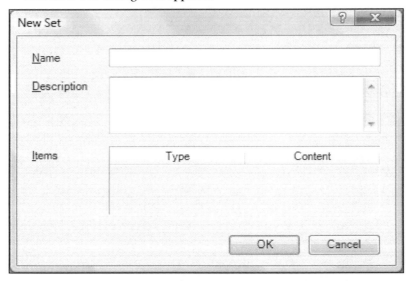

    4   Type a name (compulsory) and a description (optional), then **[OK]**.

Next, you need to define the members of your set:
    1   Select the item or items that will form a set.
    2   Go to **Create | Collections | Add To Set**
          or right-click any Project Item and select **Add To Set...**

The **Select Set** dialog box appears:

    3   Select a set and confirm with **[OK]**.

You can also select items or shortcuts from any folder and paste them into a set. When using **Find**, **Advanced Find**, or **Grouped Find**

30

the result can easily be added to a set. Sets can be used as an alternative to storing results in a subfolder to **Search Folders**:
1 Select an item (shortcut) or items (shortcuts) that will form a new set.
2 Go to **Create | Collections | Create As Set**.
The **New Set** dialog box is shown.
3 Type a name of the new set.
4 Confirm with [**OK**].

> **Sets**: Sets are a powerful organizational tool in NVivo, but beginning and intermediate users are sometimes confused by their functionality. The main function of sets is to allow users the flexibility of organizing project items into temporary or permanent groups.
>
> For example, we are involved in a project involving interview data, focus group data, writing samples, and social media data for a group of 20 undergraduate social science students. As a team, we could organize these data sources according to type of data source (e.g., an interview folder, a focus group folder, etc.) or we could organize these data sources according to student (e.g., a folder for Student 1, a folder for Student 2, etc.). While each method of organization has its merits, Sets allows us to organize project items according to type of data source AND create a Set organizing data sources according to student. As alternative methods of organizing text items present themselves, more and more sets can be generated.

Undo

The undo-function can be made in several steps backwards. Undo only works for commands made after the last save:
1 Go to **Undo** on the **Quick Access Toolbar** or [**Ctrl**] + [**Z**].

The arrow next to the undo-icon makes it possible to select which of the last five commands that shall be undone. When you select the first option only the last command is undone and when you select the last option all commands will be undone.

The option **Redo** (Undo – Undo) is available in Word but not in NVivo.

The Ribbon

Commands are organized into logical groups, collected together under tabs. Each tab relates to a type of activity, such as creating new Project Items or analyzing your source materials.

The **Home, Create, External Data, Analyze, Query, Explore, Layout,** and **View** tabs are always visible. The other tabs are 'contextual' which means that they are shown only when needed. For example, the **Picture** tab is shown only when a picture content is visible.

Within each tab, related commands are grouped together. For example, the **Format** group on the **Home** tab contains commands for setting font size, type, bold, italics and underline.

The ribbon is optimized for a screen resolution of 1280 by 1024 pixels, when the NVivo window is maximized on your screen. When the NVivo window is not maximized and the ribbon is smaller, you may not see all the icons or text.

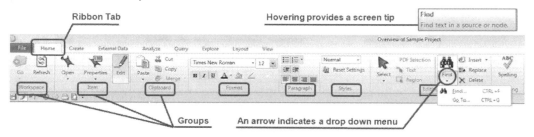

Command example: Go to **Home** | **Editing** | **Find** → **Find...** or **[Ctrl] + [F]**

The **Quick Access Toolbar** is always visible and provides quick access to frequently-used commands. By default, the Save, Edit and Undo commands are available in Quick Access Toolbar. You can customize the Quick Access Toolbar by adding or removing commands. You can also move the Quick Access Toolbar above or below the ribbon by clicking the small arrow:

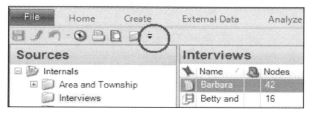

Select the options *Show Above the Ribbon* and *Minimize the Ribbon*. The menu tabs are shown again as soon as you point at any menu alternative.

The **Home** tab provides commands related to formatting (e.g., paragraph styles) and workflow (e.g., cut and paste):

The **Create** tab provides commands related to making new Project Items (e.g., creating a new Node):

The **External Data** tab provides commands related to importing and exporting Project Items:

The **Analyze** tab provides commands related to coding, linking, annotating and Framework Matrices:

The **Query** tab provides commands related to searching and querying your data.

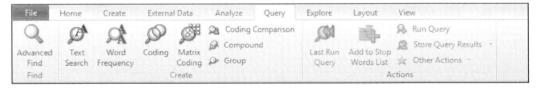

The **Explore** tab provides commands related to analytic representations.

The **Layout** tab provides commands related to list views and tables:

The **View** tab provides commands related to visual aspects of the Project Item interface (e.g., docking and coding stripes):

- ♦ -

The following tabs are context dependent, meaning they only become available depending on the Project Item type that is open:

The **Media** tab provides commands related to audio and video:

The **Picture** tab provices commands related to images:

The **Report** tab provides commands related to the Report Designer:

The **Chart** tab provides commands related to creating and modifying charts:

The **Model** tab provides commands related to creating and modifying models:

The **Cluster Analysis** tab provides commands related to conducting and formating cluster analyses:

The **Tree Map** tab provides commands related to modifying Tree Maps:

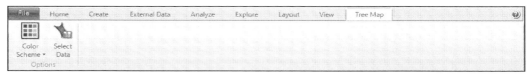

The **Word Tree** tab provides commands related to modifying Word Trees:

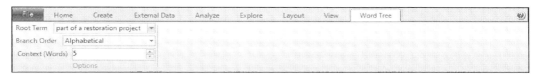

The **Graph** tab provides commands related to modifying graphs:

The guidelines of this book have adopted certain rules when referring to any command using the ribbon tabs. See under Graphic Conventions, page 16.

Application Options

NVivo project settings can be adjusted for an individual project or for the NVivo software overall. **Application Options** adjust settings for the software overall, and some changes you make only apply to new projects and will therefore have an effect on the **Project Properties** (see page 51) settings for future projects:

1 From the NVivo Welcome Screen go to **File → Options**.

Tip: Display plain text for Nodes with <value> or more sources ensures better performance for large projects. Restore source formatting by going to **View | Detail View | Node → Rich Text**.

The General Tab

The **General** tab contains default options for working with the NVivo interface such as user interface language and coding context.

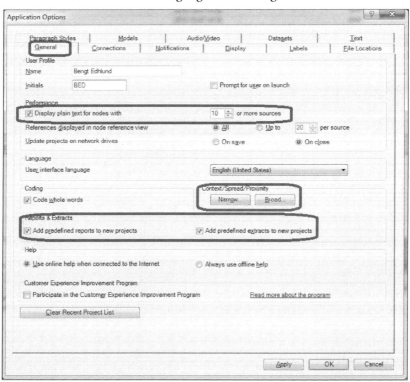

Settings made here take immediate effect in an ongoing project and will also become default for new projects. Here you can change the user interface language.

The [**Narrow...**] and [**Broad...**] buttons define principles for context/spread/proximity when viewing and searching Nodes. Each source type has its own settings: Text, Media, Transcript, Region, Log and Dataset.

Whether you want to inherit predefined reports and extracts is also determined here.

The Connections Tab

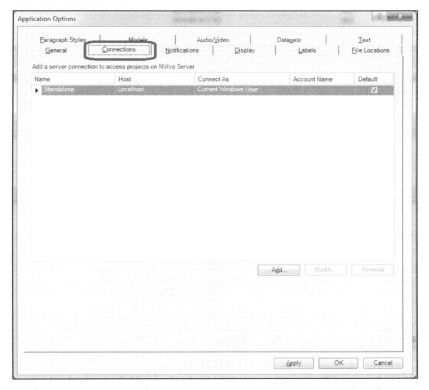

The **Connections** tab contains settings pertaining to NVivo Server – additional QSR proprietary teamwork software (see page 295).

> **Tip:** Why take a chance on losing valuable work? We recommend that a save reminder displays every 10 minutes, instead of the default 15.

The Notifications Tab
The **Notifications** tab contains default options for save reminders and software update checks.

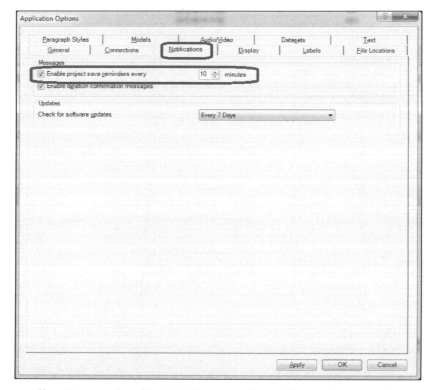

All settings under this tab take immediate effect on an ongoing project.

The Display Tab

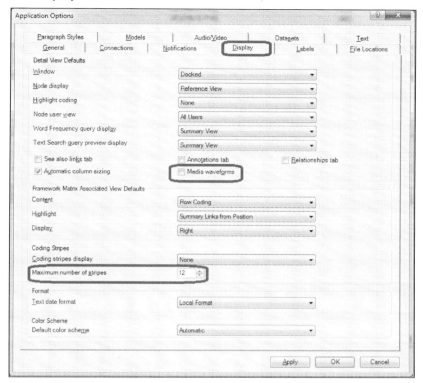

The Display tab contains default options for visual cues in NVivo such as coding stripes, highlighting, and tabs.

We suggest unchecking the display of Media waveforms as the waveform often disturbs other graphic information like coding stripes, links and selections.

We also suggest increasing the maximum number of coding stripes beyond the default number which is 7. The number of stripes cannot be extended beyond 200.

In case you prefer always to open your windows undocked, then set Detail View as Window *Floating*.

For settings related to Framework Matrix, see page 233.

All settings under this tab take immediate effect on an ongoing project.

The Labels Tab

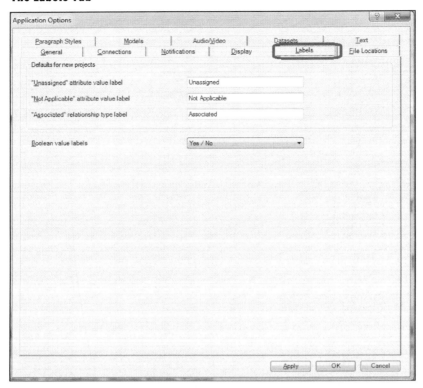

The **Labels** tab allows you to customize the names of attributes, values and the Associated relationship type.

Settings made here will take effect next time a project is created. If you want to make changes in the current project, use **Labels** tab in **Project Properties** dialog box (see page **Error! Bookmark not defined.**).

The File Locations Tab

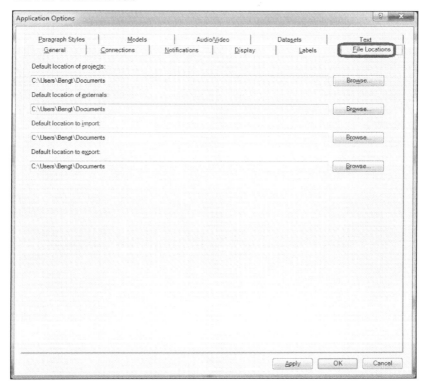

The **File Locations** tab contains default file locations of projects, externals and the default locations of imported and exported Project Items and data.

All settings under this tab take immediate effect on an ongoing project.

The Paragraph Styles Tab

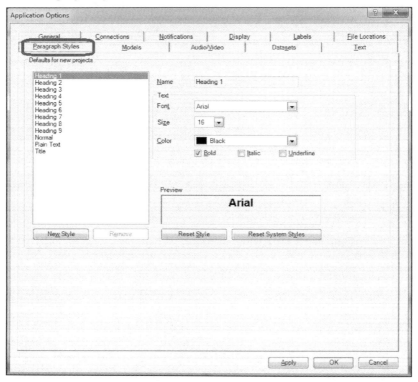

The **Paragraph Styles** tab contains default options for NVivo styles (see page 72). Changes made under Application Options will be available next time a new project is created. Existing project settings can be modified under the **Paragraph Styles** tab in **Project Properties** (see page 55).

The Model Styles Tab

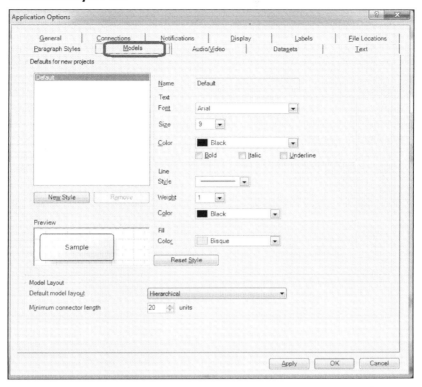

The **Model Styles** tab makes it possible to define a model's fonts, line weight, style, color, and fill color for shapes. You can create any set of new model styles for existing projects. Changes made under Application Options will be available next time a new project is created. Existing project settings can be modified under the **Model Styles** tab in **Project Properties** (see page 56).

> **Tip:** Settings for the threshold value of embedded audio and video files are set under: *Embed media in project if file size less than <value> MB*. Max is 40 MB for single users. Embedded or not, you can always code, link and create transcript rows in a media item.

The Audio/Video Tab

The **Audio/Video** tab contains settings for the skip interval for skipping forward and skipping backward. The threshold value for embedding is set here. These settings have an immediate effect on an open project. You can also create custom transcript fields (or columns) for audio and video items. However, these fields will come into effect for new projects. To create custom transcript fields in an existing project go to **File → Project Properties**, and select the **Audio/Video** tab, see page 57).

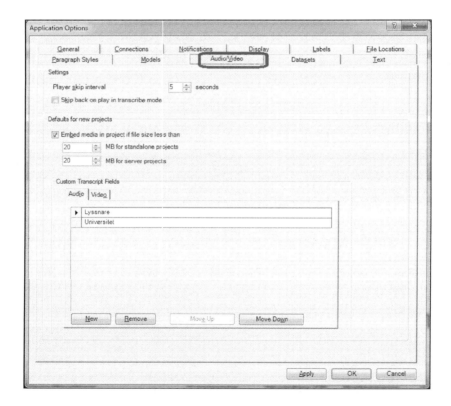

The Datasets Tab

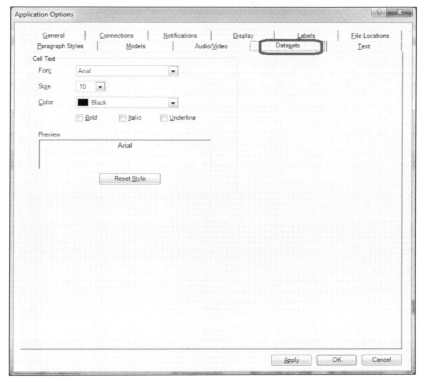

The **Datasets** tab allows you to adjust the font, size and color of cell text. Modifications made here will take effect next time a Dataset is opened.

The Text Tab

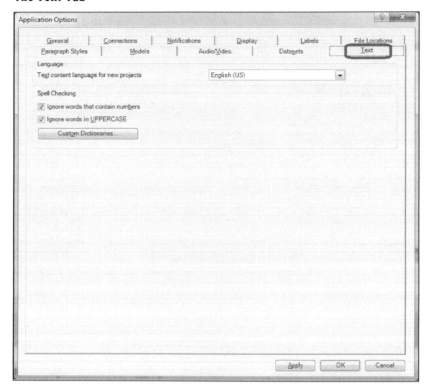

The **Text** tab allows you to make settings for the content language and the spell checking dictionaries that you want to use. The setting of language does not have an effect on the ongoing project. For the next new project the chosen language will be default. If you want to change content language for the current project go to the **General** tab of **Project Properties dialog box** (see page 51).

The supported languages are: Chinese (PRC), English (UK), English (US), French, German, Japanese, Portuguese and Spanish. If you use any other language then you can set the language as *Other*.

The button **[Custom Dictionaries...]** can be used to appoint one specific folder for each language which can include the custom dictionary, **<filename>.DIC**. This is a normal text-file and can be opened and edited with Notepad. If you already have a custom dictionary from before you can name it **<filename>.DIC** and store it in the defined folder. Even the setting *Other* can have its own custom dictionary.

Alternate Screen Layouts

Sharing the screen with four windows can sometimes make reading Area 4 difficult. But NVivo offers an alternate screen layout to split the screen space vertically instead of horizontally between Area 3 and Area 4.

    1   Go to **View | Workspace | Detail View → Right**.

The Right Detail View is very handy when coding with drag-and-drop (see page 154).

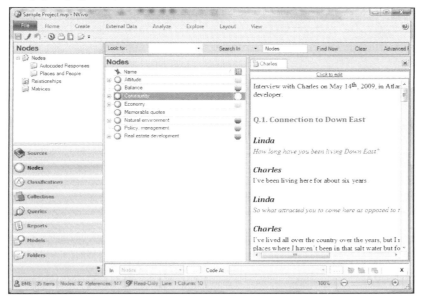

To revert to original setting of the screen:

    1   Go to **View | Workspace | Detail View → Bottom**.

New to NVivo 10, your **Detail View** setting preference can now be saved in between sessions.

For more screen space it is also possible to temporarily close Areas 1 and 2.

1. Go to **View | Workspace → Navigation View** or [**Alt**] + [**F1**], which is a toggling function.

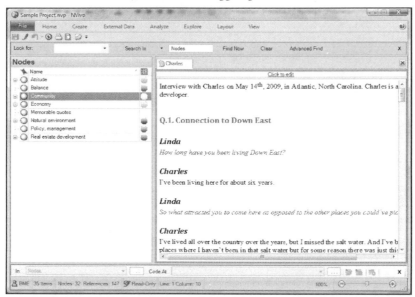

This setting is saved during current session even if you open another project. You can navigate to other folders by going to **Home | Workspace | Go key** → <select> or [**Ctrl**] + [**1 – 8**].

3. BEGINNING YOUR PROJECT

An NVivo project is a term used for all source documents and other items that altogether form a qualitative study. A project is also a computer file that houses all those Project Items.

NVivo can only open and process one project at a time. It is however possible to start the program twice and open one project in each program window. Cut, copy, and paste between two such program windows is limited to text, graphics and images and not Project Items like documents or Nodes.

A project is built up of several items with different properties. There are internal sources (i.e., documents, memos), external sources (i.e., web sites), Nodes and queries.

Creating a New Project

The Welcome screen will greet you each time you launch NVivo, and it is from here that you have the option to create a new project file:

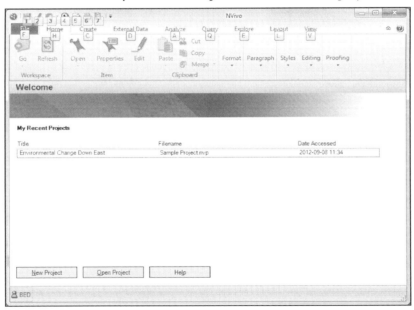

Your most recent projects are listed on the Welcome screen. The [**New Project**] button makes it possible to create a new project. *Alternatively,* you can also create a new project while navigating inside an existing project:

1   Go to **File → New**
    or [**Ctrl**] + [**N**]

    or the icon   on the Quick Access Toolbar.

The **New Project** dialog box appears:

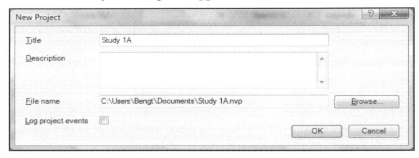

You must type a name for your project file, but the description is optional. The file path for your project is seen in the **File name** box. Click **[Browse]** to select a new location for your project file. The default location of the project file is determined in the **Application Options** dialog box, under the **File Locations** tab (see page 41). The name of a project can later be changed without changing the file name. Any open NVivo projects will be closed as NVivo opens a new or an existing project.

Sources & Project Size

NVivo is capable of importing and creating a variety of file types as data (e.g., text sources, tables, images, video, PDFs, etc.) Collectively, these items care called *Sources*. We'll discuss sources at length over the next few chapters, but for now it's important you understand that NVivo can either *import* Sources to a project file or *link* Sources externally to a Project file.

Files that are imported into NVivo are amalgamated by the software, which means that they become a part of the NVivo project. These are called *Internal Sources*. For example, any changes you make to imported sources (e.g., a text Source) are not reflected in the original document (e.g., a Microsoft Word text file).

Files that are linked into NVivo are only referenced by the software, which means that they exist independently of the NVivo project. These are called *External Sources* and cannot be coded, only the external item (the text inside NVivo) can be coded.

Audio- and videofiles are special as they can either be embedded or remain stored outside NVivo. If such files remain outside NVivo they can still be handled as if they were embedded, that is you can link and code, etc. Therefore, even not embedded media files are *Internal Sources*. It is the size of such files that decides if it should be embedded or not. A threshold value set by the user decides. The threshold value however, cannot exceed 40 MB (read more on page 94).

An NVivo 10 Project file size is maximum 10 GB. Bear in mind that large Project Items (e.g., audio and video files) can be stored outside the project file. Linking to external files allows you to keep the project file size down. Using NVivo Server (see Chapter 22, Collaborating with NVivo 10) allows for a maximum project file size of 100GB.

Project Properties

When a new project is created some settings from the **Application Options** dialog box are inherited. This dialog box opens by going to **File → Options**, and the settings that are inherited are found under these tabs: **Labels**, **Paragraph Styles, Model Styles**, and **Audio/Video**. Modifications and templates which are made in the **Project Properties** dialog box are only valid for your current project:

1 Go to **File → Info → Project Properties**.

The General Tab

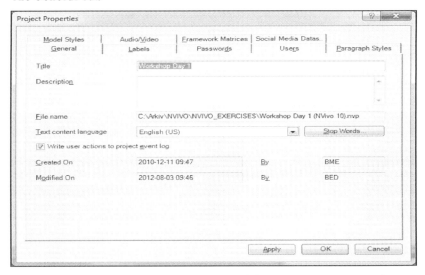

In this dialog box it is possible to modify your project name, but not the NVivo project file name. From the **Text content language** drop-down list you will, if available, select the language of the data used in the project, otherwise select *English* or *Other*. Your content language will be the default language for spell check, as well as an important setting for Text Search Queries and Word Frequency Queries. For all languages except *Other* a default stop word list is built in. The stop words list can be edited using the **[Stop Words]** button or while using Word Frequency Queries (see Chapter 13, Queries). Such customized stop word lists are only valid for the current project. Even when the content

Tip: Your stop words list can be edited with the button **[Stop Words]**. Remember, customized stop words are only valid for the current project.

language setting is *Other* you can build a customized stop word list.

The Description (max 512 characters) can be modified. *Write user actions to project event log* is optional. When activated you can open this log with:

File → Info → Open Project Event Log

or delete it with:

File → Info → Clear Project Event Log

The Labels Tab

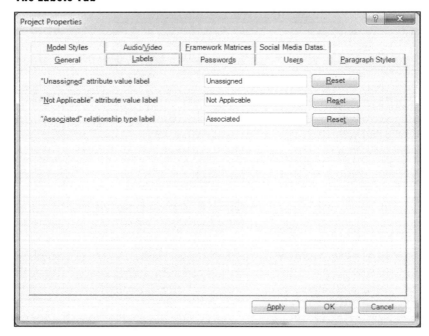

Under the **Labels** tab you can change some of your project's 'labels'. The **[Reset]** buttons reset to the values defined in the **Application Options** dialog box, under the **Labels** tab (see page 40).

The Passwords Tab

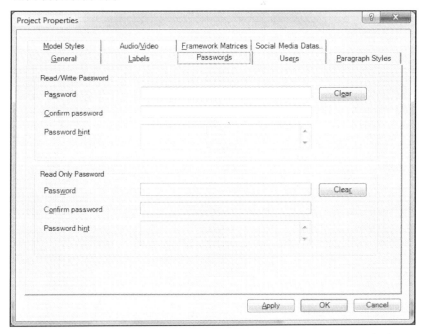

Under the **Passwords** tab you can define separate passwords for opening and editing your current project.

The Users Tab

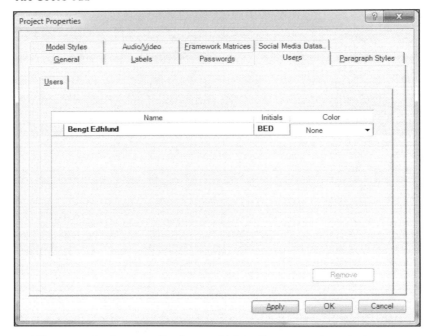

All users who have actively worked in the current project are listed here. The current user is identified by bold letters. You can replace a user with someone else on the list by selecting the user who shall be replaced (triangle) and using the [**Remove**] button. Select who will replace the deleted user by selecting from the list of users.

Users can also be given an individual color marking. Use the drop-down list in the Color column and select color. This color marking can be used when viewing coding stripes per user.

The Paragraph Styles Tab

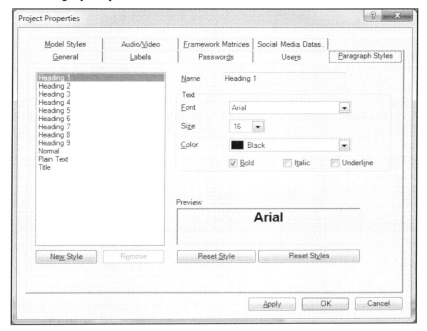

Under the **Paragraph Styles** tab, you can redefine your paragraph styles. The [**Reset Styles**] buttons reset to values defined in the **Application Options** dialog box, (see page 42).

The Model Styles Tab

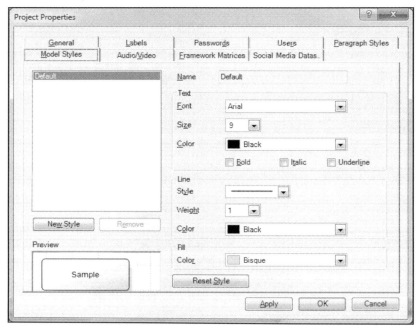

When a new project is created, NVivo retains the styles that were previously defined with the **Application Options** dialog box, under the **Model Styles** tab (see page 43). New styles created in the **Project Properties** dialog box are only valid for the current project. Using [**Reset Style**] brings you back to settings made in **Application Options**.

The Audio/Video Tab

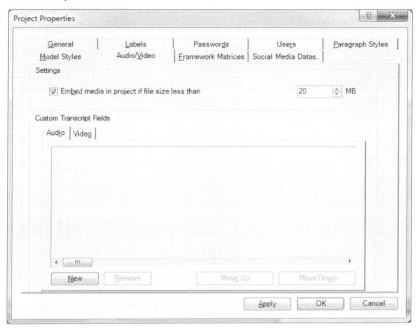

The settings for new projects are inherited from the **Application Options** dialog box, the **Audio/Video** tab (see page 44). Modifications made here are only valid for the current project.

When you need to create Custom Transcript Fields in your current project then you may use this dialog box. The [**New**] button defines more fields like Speaker, Affiliation with separate fields for Audio and Video.

The Framework Matrices Tab

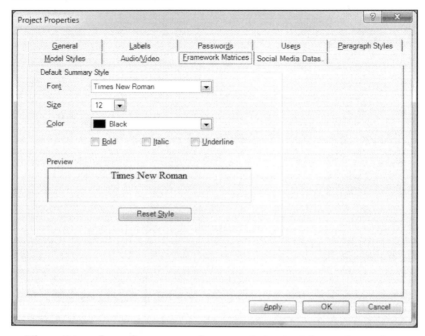

The settings here for new projects are inherited from the style Normal in the **Application Options** dialog box, the **Paragraph Style** tab (see page **Error! Bookmark not defined.**). Modifications made here are only valid for the Framework Matrices summaries in the current project.

The Social Media Dataset Tab

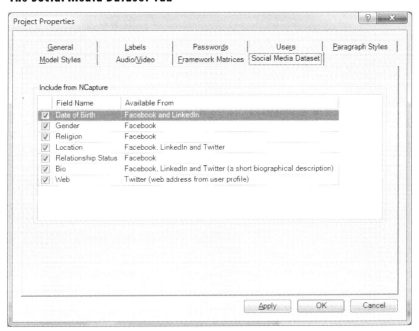

New to NVivo 10, social media Datasets can be imported via NCapture files containing Facebook, Twitter or LinkedIn data. This tab allows you to toggle the types of data you wish to capture from each social networking site.

Merging Projects

Projects can be merged by importing one project to another:
1. Open the project into which you wish to import a project.
2. Go to **External Data | Import | Project**.

The **Import Project** dialog box appears:

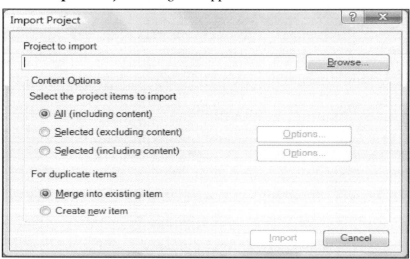

3. The [**Browse...**] button opens a file browser. Search for the project file to be imported.
4. Select the item options that you need for the import.
5. Confirm with [**Import**].

An **Import Project Report** is now shown listing all items.

Tip: For users who don't have access to NVivo Server, merging projects is a useful function for teams collaborating on the same project. Users can independently make changes to their project and, later, import their changes into a project 'master' file.

Exporting Project Data

All Project Items (except folders) can be exported in various file formats. For example, project Memos can be created in NVivo and then exported as .doc files so they can be shared with collaborated over email:
1. Open a project.
2. Go to **External Data | Export | Export → Project**.

The **Export Project Data** dialog box appears:

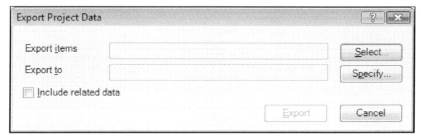

At **Export items** and the **[Select]** button you decide what items that shall be exported and at **Export to** and the **[Specify]** button you decide the name and location of the exported project data.

Exporting Project Items

All Project Items (except folders) can be exported in various file formats:

1. Select the item or items that you want to export, for example two Nodes.
2. Go to **External Data | Export | Export → Export Node...** or right-click and select **Export → Export Node...** or **[Ctrl] + [Shift] + [E]**.

The **Export Options** dialog box appears:

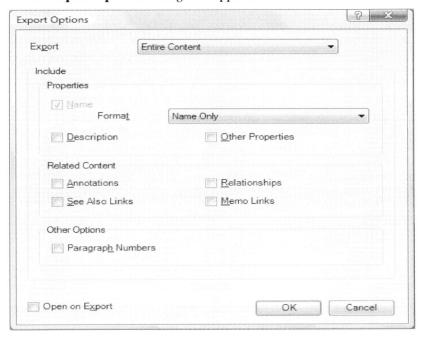

3   Select the options for the export and decide file name, file location, and file type, then [OK].

The option *Entire Content* creates a HTML-page with several files and folders that can be uploaded to a Web-server.

Save and Security Backup

You can save the project file at any time during a work session. The complete project is saved; it is not possible to save single Project Items.

1   Go to **File → Save**
    or **[Ctrl] + [S]**

    or the icon 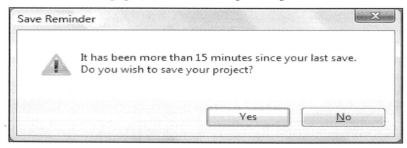 on the Quick Access Toolbar.

If the option *Enable project save reminders every 15 minutes* has been chosen (see page 38) the following message will show:

2   Confirming with **[Yes]** saves the whole project file.

Securely backup of your NVivo project is easy since the whole project is one file and not a structure of files and folders. Use Windows native tools for backup copies and follow the backup routines that your organization applies. The command **File → Manage → Copy Project** creates a copy of the project at the location that you decide while the current project remains.

4. HANDLING TEXT SOURCES
Documents
From interview transcripts to government white papers, text data makes up the majority of qualitative research data. Text items can be easily imported from files created outside NVivo, like Word documents or text notes from Evernote. Text items can also be created by NVivo as most word processing tools and functions are incorporated in NVivo software, which we'll discuss in the next chapter.

Importing Documents

This section is about text-based sources that can be imported and these file types are: .DOC, .DOCX, .RTF, .TXT, and text-only Evernote export files (.ENEX). When text files are imported into NVivo, they become Project Items within the Source folder:

1 Go to **External Data | Import | Documents**
 Default folder is **Internals**.
 Go to 5.

alternatively

1 Click on [**Sources**] in Area 1.
2 Select the **Internals** folder in Area 2 or its subfolder.
3 Go to **External Data | Import | Documents**.
 Go to 5.

alternatively

3 Click on any empty space in Area 3.
4 Right-click and select **Import Internals → Import Documents...**
 or [**Ctrl**] + [**Shift**] + [**I**].

alternatively

3 Drag and drop your file's icon from an outside folder into Area 3.
 Go to 5.

In each case, the **Import Internals** dialog box is shown:

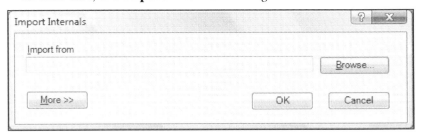

5   The [**Browse...**] button gives access to a file browser and you can select one or several documents for a batch import. To select multiple documents use [**Shift**] + left click.
6   When the documents have been selected, confirm with [**Open**].

The [**More** >>] button offers several options:

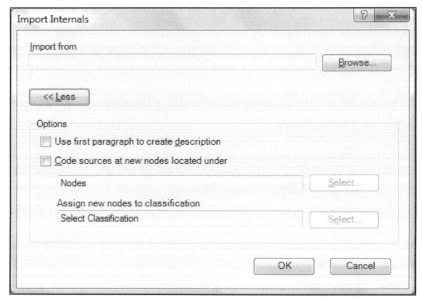

Use first paragraph to create descriptions. NVivo copies the first paragraph of the document and pastes it into the Description field.

Code sources at new Nodes located under. Each Source Item will be coded at a Node (typically a Case Node) with the same name as the imported file and located in a folder or under a parent Node that has been selected. Also you must assign the Nodes to a Classification when importing (see Chapter 11, Classifications).

7   Confirm the import with [**OK**].

When only *one* document has been imported, the **Document Properties** dialog box appears:

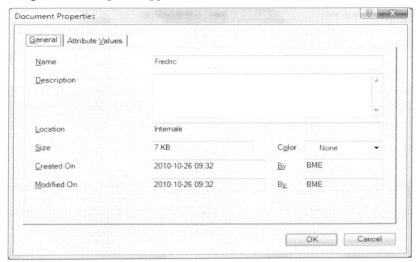

Within this dialog box you can modify the name of the Source Item and optionally add a description.

8 Confirm with [**OK**].

Creating a New Document

You can also create your own text items within NVivo, much the same as creating a Word document or text note in Evernote.

1 Go to **Create | Sources | Document**
   Default folder is **Internals**.
   Go to 5.

alternatively

1 Click on [**Sources**] in Area 1.
2 Select the **Internals** folder in Area 2 or its subfolder.
3 Go to **Create | Sources | Document**
   Go to 5.

alternatively

3 Click on any empty space in Area 3.
4 Right-click and select **New Internal → New Document...**
   or [**Ctrl**] + [**Shift**] + [**N**].

The **New Document** dialog box appears:

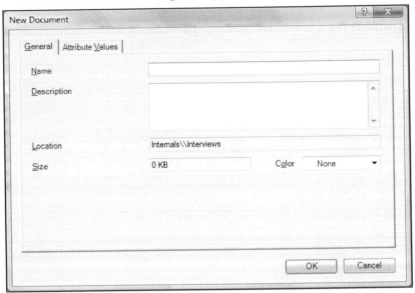

5   Type a name (compulsory) and a description (optionally), then **[OK]**.

Here is a typical list view in Area 3 of some Source Items:

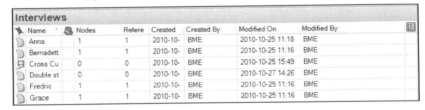

Opening a Document

Now that you have imported or created a list of Source Items, you can easily open one or more items anytime you see fit:

1   Click on **[Sources]** in Area 1.
2   Select the **Internals** folder in Area 2 or its subfolder.
3   Select the document in Area 3 that you want to open.
4   Go to **Home | Item | Open**
     or right-click and select **Open Document...**
     or double-click on the document in Area 3
     or **[Ctrl]** + **[Shift]** + **[O]**.

Please note, NVivo only allows you to open one document at a time, but several documents can stay open simultaneously.

Exporting Documents

As mentioned, you may wish at some point to export a text Source Item, such as a Memo you wrote inside NVivo but now need to email to a collaborator.

1. Click [**Sources**] in Area 1.
2. Select the **Internals, Externals** or **Memos** folder in Area 2 or its subfolder.
3. Select the document or documents in Area 3 that you want to export.
4. Go to **External Data | Export | Export → Export Document...**
   or right-click and select **Export → Export Document...**
   or [**Ctrl**] + [**Shift**] + [**E**].

The **Export Options** dialog box appears:

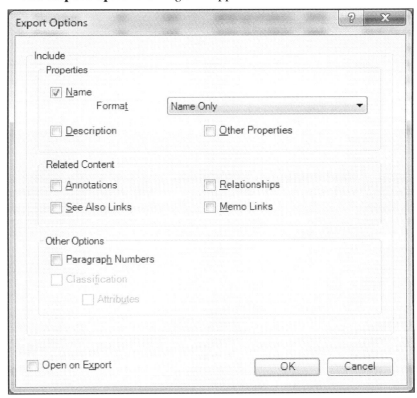

5. Select the options that you want. Confirm with [**OK**].
6. Decide file name, file location, and file type. Possible file types are: .DOCX, .DOC, .RTF, .TXT, .PDF, or .HTML. Confirm with [**Save**].

Remember, coding made on text items cannot be transferred when a Source Item is exported.

External Items

For any number of reasons, you may wish to refer to external items outside of your NVivo project (i.e., a web site, a file too large or a file type that is incompatible). NVivo allows you to create external items that can act as placeholders or links.

Creating an External Item
1. Go to **Create | Sources | External**
   Deafault folder is **Externals**.
   Go to 5.

alternatively
1. Click on **[Sources]** in Area 1.
2. Select the **Externals** folder in Area 2 or its subfolder.
3. Go to **Create | Sources | External**.
   Go to 5.

alternatively
3. Click on any empty space in Area 3.
4. Right-click and select **New External...**
   or **[Ctrl] + [Shift] + [N]**.

The **New External** dialog box appears:

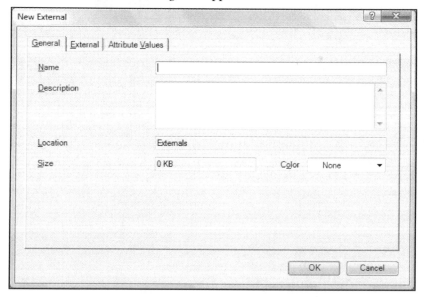

5. Type name (compulsory) and description (optional), then go to the **External** tab.

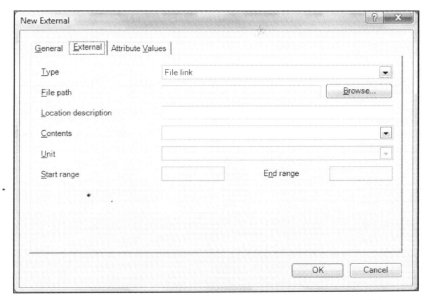

6   At **Type** select *File link* and then use the **[Browse...]** button to find the target file. Alternatively, at Type select *Web link* and type or paste the URL in the text box below.

7   Confirm with **[OK]**.

This is a typical list view in Area 3 of some external items:

Opening an External Item

External items act identical to internal items within NVivo's Sources folder: they can contain text and that text can be edited and coded. To open an external item for viewing or editing:

1   Click on **[Sources]** in Area 1.
2   Select the **Externals** folder in Area 2 or its subfolder.
3   Select the external item in Area 3 that you want to open.
4   Go to **Home | Item | Open**
    or right-click and select **Open External...**
    or double-click on the external item in Area 3
    or **[Ctrl] + [Shift] + [O]**.

Remember, NVivo can only open one external item at a time, but several items can stay open simultaneously.

Opening an External Source
Unlike internal items, external items are necessarily linked to external sources, which can be opened through NVivo:
1. Click on **[Sources]** in Area 1.
2. Select the **Externals** folder in Area 2 or its subfolder.
3. Select the external item in Area 3 that has a link to the external file or URL that you want to open.
4. Go to **External Data | Files | Open External File** or right-click and select **Open External File**.

Editing an External Source or Link
1. Click on **[Sources]** in Area 1.
2. Select the **Externals** folder in Area 2 or its subfolder.
3. Select the external item in Area 3 that you want to edit.
4. Go to **Home | Item | Properties** or right-click and select **External Properties...** or **[Ctrl] + [Shift] + [P]**.

The **External Properties** dialog box appears.

5. Select the **External** tab and if you want to link to a new target file use **[Browse...]**. If you want to modify a web link change the URL.

Exporting an External Item
Similar to internal items, external items can be exported. However, the linked external file or the web link is not included in the exported item, only the external item text contents are exported.
1. Click on **[Sources]** in Area 1.
2. Select the **Externals** folder in Area 2 or its subfolder.
3. Select the external item or items in Area 3 that you want to export.
4. Go to **External Data | Export | Export** or right-click and select **Export → Export External...** or **[Ctrl] + [Shift] + [E]**.

The **Export Options** dialog box appears.

5. Select the options that you want. Confirm with **[OK]**.
6. Decide file name, file location, and file type. Possible file types are: .DOCX, .DOC, .RTF, .TXT, .PDF, or .HTML. Confirm with **[Save]**.

5. EDITING TEXT IN NVIVO

Whether you import a text document or create a new one, NVivo 10 contains most of the functions of modern word processing software. Notwithstanding the fact that text document files are often imported, understanding how to edit text in NVivo is useful. Aside from its ability to edit existing source documents, you can use NVivo's word processing functionality to compose Memos, Externals, and Framework Matrix summaries.

Formatting Text

Remember, each time a Source Item is opened it is Read-Only. Therefore, click *Click to edit* (or **Home | Item | Edit** or **[Ctrl] + [E]**) at the top of a Source Item before editing.

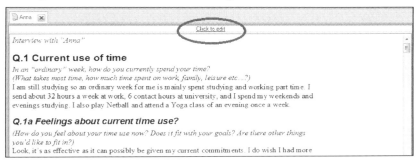

Selecting the whole document:
1. Position the cursor anywhere in the document.
2. Go to **Home | Editing | Select → Select All** or **[Ctrl] + [A]**.

> **Tip: Selecting Text**
> Select a passage of text by holding left-click and mousing over it. Double left-clicking on a single word highlights just that word. And did you know that triple left-clicking on a single word selects the whole paragraph? Both of these shortcuts can be when coding.

Changing Fonts, Font Style, Size, and Color
1. Select the text you want to format.
2. Go to **Home | Format → Font...**

The **Font** dialog box appears:

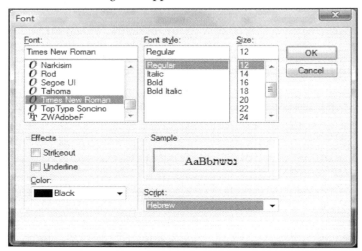

3. Select the options you need and confirm with **[OK]**.

Selecting a Style
1. Position the cursor in the paragraph you want to format.
2. Go to **Home | Styles**.
3. Select from the list of styles.
4. Confirm with **[OK]**.

Resetting to previous style is possible as long as the project has not been saved after the last change:
1. Position the cursor in the paragraph you want to reset.
2. Go to **Home | Styles | Reset Settings**.

Aligning Paragraphs

Selecting Alignments
1. Position the cursor in the paragraph you want to format.
2. Go to **Home | Paragraph**.
3. Select from the list of alignment options.

Selecting Indentation
1. Position the cursor in the paragraph for which you want to change the indentation.
2. Go to **Home | Paragraph**.
3. Select increased or decreased indentation.

Creating Lists
1. Select the paragraphs that you want to make as a list.
2. Go to **Home | Paragraph**.
3. Select a bulleted or numbered list.

Finding, Replacing and Navigating Text

Finding Text
1. Open a document.
2. Go to **Home | Editing | Find → Find...**
   or **[Ctrl] + [F]**.

The **Find Content** dialog box appears:

3. Type a search word, then click **[Find Next]**.

The **Style** option makes it possible to limit the search from *Any* to a certain style.

Please note the option *Match case* which makes it possible to exactly match *UPPERCASE* or *lowercase* and the option *Find whole word* which switches off the free text search.

Searching and Replacing

1. Open a document.
2. Go to **Home | Editing | Replace**
   or **[Ctrl] + [H]**.

The **Replace Content** dialog box appears:

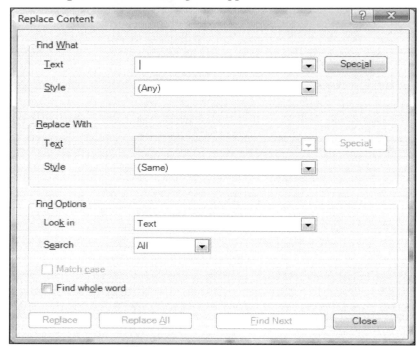

3. Type a find word and a replace word, then **[Replace]** or **[Replace All]**.

The option Style near **Find What** makes it possible limit the search from All to any given style and the option Style near **Replace With** makes it possible to replace the found word as well as change the style from Same to any given style.

Please notice the option *Match case* which makes it possible to exactly match *UPPERCASE* and *lowercase* and the option *Find whole word* which switches off the double-sided auto truncation.

Spell Checking

NVivo comes with built-in dictionaries for English (UK), English (US), French, German, Portuguese and Spanish. If your source materials use specialized terms or abbreviations that are not in the built-in dictionary, you can add these words to a custom dictionary. Each of these languages can have its own custom dictionary.

When you spell check a source, NVivo flags words that are not in the built-in or custom dictionary. You can decide whether you want to ignore flagged words, correct them or add them to the custom dictionary.

You can spell check source content when the source is open in edit mode. You can spell check:
- Documents
- Memos
- Audio and video transcripts (the Content column only)
- Picture logs (the Content column only)
- Framework Matrices
- Externals

You can also spell check Annotations in any type of source, including non-editable source types such as Datasets and PDFs. You can also spell check annotations displayed in Node Detail View.

You can set your spell check preferences in NVivo's application options—for example, you can choose whether or not to flag all uppercase words (e.g. USA) as spelling mistakes.

1 Open a Source Item in Edit mode.
2 Go to **Home | Proofing | Spelling**
   or [**F7**].

The **Spelling:** <**Language**> dialog box appears:

The meanings of these buttons are:

[Ignore Once]	Ignore and move to next
[Ignore All]	Ignore all instances in the whole source and move to next
[Add To Dictionary]	The word is added to the custom dictionary and will not be flagged from now on
[Change]	Changes the spelling to the highlighted suggested word
[Change All]	Changes the spelling to the highlighted suggested word at all instances in the whole source and move to next
[Cancel]	Stops the spell checking

When you want to spell check any Annotation, open the annotations window and keep the cursor within the annotation. If you have more than one annotation in the same source the spell checker will run through all of them. The source itself need not be in edited mode when you spell check an annotation.

For more on settings for content languages and dictionaries (see page 46).

Selecting Text

Selecting text: Click and drag
Selecting one word: Double-click
Selecting a paragraph:
   1   Position the cursor in the paragraph you want to select.
   2   Go to **Home | Editing | Select → Select Paragraph**
       or triple-click.

Selecting the whole document:
1. Position the cursor anywhere in the document.
2. Go to **Home | Editing | Select→ Select All** or **[Ctrl] + [A]**.

> 'Go To' options vary depending on the source type (e.g., Documents, PDFs, Datasets, Pictures, and Audio or Video). Possible Go To options include Paragraph, Character Position, See Also Link and Annotation (above), as well as Dataset Record ID, Log Row, Page, Source, Time, and Transcript Row.

Go to a Certain Location

1. Go to **Home | Editing | Find → Go to...** or **[Ctrl] + [G]**.

The **Go to** dialog box appears:

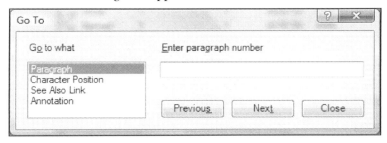

2. Select option at **Go to what** and when required, a value.
3. Click on **[Previous]** or **[Next]**.

Creating a Table

1. Position the cursor where you want to create a table.
2. Go to **Home | Editing | Insert → Insert Text Table...**

The **Insert Text Table** dialog box appears:

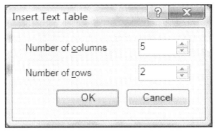

3. Select number of columns and number of rows in the table.
4. Confirm with **[OK]**.

Inserting Page Breaks, Images, Dates, and Symbols

Inserting a Page Break
1. Position the cursor where you want to insert a page break.
2. Go to **Home | Editing | Insert → Insert Page Break**.

A page break is indicated with a dotted line on the screen.

Inserting an Image
1. Position the cursor where you want to insert an image.
2. Go to **Home | Editing | Insert → Insert Image...**
3. Select an image with the file browser. Only .BMP, .JPG and .GIF file formats can be inserted.
4. Confirm with **[Open]**.

Inserting Date and Time
1. Position the cursor where you want to insert date and time.
2. Go to **Home | Editing | Insert → Insert Date/Time** or **[Ctrl] + [Shift] + [T]**.

Inserting a Symbol
1. Position the cursor where you want to insert a symbol.
2. Go to **Home | Editing | Insert → Insert Symbol** or **[Ctrl] + [Shift] + [Y]**
3. Select a symbol from the **Insert Symbol** dialog box, confirm with **[Insert]**.

Zooming

1. Open a document.
2. Go to **View | Zoom | Zoom | Zoom...**

The **Zoom** dialog box appears:

Tip: Our preferred method of zooming in NVivo is **[Ctrl]** + mouse wheel. **[Ctrl]** + moving the mouse wheel forward allows zooming in; **[Ctrl]** + moving the mouse wheel backward allows zooming out.

3. Select a certain magnification and confirm with **[OK]**.

Alternatively, you may also use the Zoom-slider in the status bar below on the screen.

Alternatively, **[Ctrl]** + your mouse wheel allows zooming in or out.

You can also zoom in or out in predetermined steps:
1. Open a document.
2. Go to **View | Zoom | Zoom | Zoom In** or **View | Zoom | Zoom | Zoom Out**.

Print Previewing

1. Open a document.
2. Go to **File → Print → Print Preview**.

The **Print Options** dialog box appears:

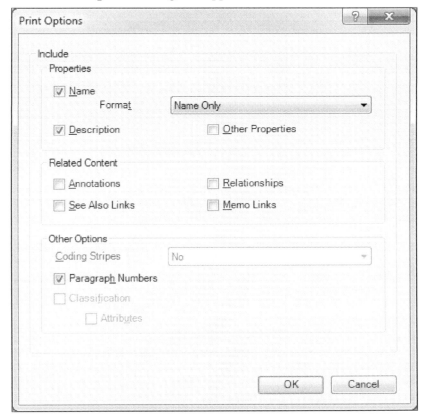

3. Select options for the preview.
4. Confirm with **[OK]**.

As you can see from the dialog box we have selected the options Name, Description and Paragraph Numbers. This can be of great importance when working in a team. Also the page breaks are shown here and they are not on the screen.

The result can look like this:

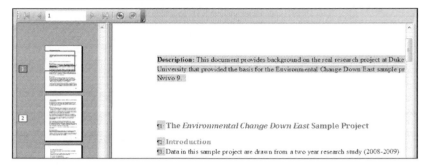

In the Print Preview window there are numerous possibilities to navigate, zoom, and change the view. The thumbnails can be hidden with **View → Thumbnails** which is a toggling function. Print all pages with **File → Print** or [**Ctrl**] + [**P**].

Printing a Document

1. Open a document.
2. Go to **File → Print → Print...** or [**Ctrl**] + [**P**].
3. The **Print Options** dialog box (same as above) now shows. Select options for the printout.
4. Confirm with [**OK**].

Printing with Coding Stripes

When you need to print a document with coding stripes (see page 167), you must first display the coding stripes on the screen. Then you need to select the option *Coding Stripes* in the **Print Options** dialog box:

The print options are: *Print on Same Page.*

Or *Print on Adjacent Pages.*

Page Setup

1. Open a document and open **Print Preview**.
2. Go to **Layout | Page | Page Setup**.

alternatively

1. Open a document.
2. Open **Print Preview**.
3. Go to **File → Page Setup...**

The **Page Setup** dialog box appears:

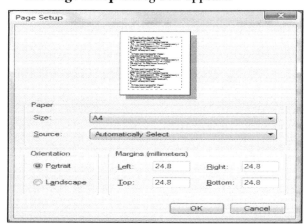

4. Decide the settings for paper size, orientation and margins, then **[OK]**.

> **Tip:** NVivo is powerful software for organizing and analyzing text documents, but it is weak as a standalone word processor. A best practice we recommend is creating a document in Word (or your preferref word processing software) and then importing the text into NVivo.

> **Tip: Make it a PDF!**
> In the event you find NVivo is mishandling your document formatting, try converting your text document to a PDF. NVivo is also a powerful tool for handling PDF files with special formatting, like multiple columns. What about if you want to import a PowerPoint presentation into NVivo? Make it a PDF!

Limitations in Editing Documents in NVivo

NVivo has certain limitations in creating advanced formatted documents.

Some of these limitations are:
- NVivo cannot merge two documents by any other means than copying/cutting and pasting text.
- It is difficult to format an image (change size, orientation, move).
- It is difficult to format a table.
- It is difficult to format a paragraph (hanging indent, first line different, line spacing).
- Copying from a Word document to NVivo loses some paragraph formatting.
- Footnotes in a Word document are lost after importing to NVivo. Word footnotes can however be manually replaced by NVivo Annotations (see page 125).
- Field codes do not exist in NVivo and these are converted to text after importing to NVivo.
- NVivo cannot apply several columns, except when used in a table. When a multi-column document is imported it is displayed on the screen as single column. The multi-column design is restored when such document is exported or printed.

Often it is better to create a document in Word and then import to NVivo. Word footnotes may however be replaced by NVivo Annotations.

Tip: Formatting your Word documents for NVivo:
1. Give your Word documents meaningful file-names. If you write an interview per document, it is advantageous if the file-name is the name of the interviewee (real name or a code name). After importing to NVivo, both the Source Item and the Case Node will be given this name. Put all interviews of same kind in the same folder, and consider the sort order. If you are using numbers in the file names then you should apply a similar series of names, with the same number of characters, like 001, 002, .. 011, 012, .. 101, 102, etc.
2. Use Word's paragraph styles to enable autocoding. For structured interviews you should create document templates with subject headings and paragraph styles.
3. Divide the text into logical, appropriate paragraphs using the hard carriage return (ENTER on your keyboard). This facilitates the coding that can take place based on a keyword and the command 'Spread Coding to Surrounding Paragraph'. Remember tripple-clicking!

6. HANDLING PDF SOURCES

Of particular interest to researchers who are conducting literature reviews, PDF documents will retain the original layout after import to NVivo and appear exactly as they were opened in Acrobat Reader. These PDFs can be coded, linked and searched as any other Source Item. One limitation is that PDF text cannot be edited nor can hyperlinks be created. Hyperlinks made in the original PDF, however, will function normally in NVivo.

Apart from bibliographic data with PDF articles downloaded from EndNote, new to NVivo 10, web pages and Evernote files can now be imported into NVivo as PDF sources. This new feature allows web pages and Evernote files to be organized, coded and even queried the same as any imported .PDF file (See Chapters 15, 18 and 19).

Importing PDF Files

1. Go to **External Data | Import | PDFs**
   Default folder is **Internals**.
   Go to 5.

alternatively

1. Click on **[Sources]** in Area 1.
2. Select the **Internals** folder in Area 2 or its subfolder.
3. Go to **External Data | Import | PDFs**.
   Go to 5.

alternatively

3. Click on any empty space in Area 3.
4. Right-click and select **Import Internals → Import PDFs...** or **[Ctrl] + [Shift] + [I]**.

alternatively

3. Drag and drop your file's icon from an outside folder into Area 3.
   Go to 5.

In each case, the **Import Internals** dialog box is shown:

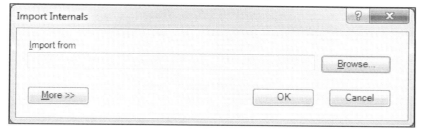

5. The **[Browse]** button gives access to a file-browser and you can select one or several PDFs for a batch import.
6. When the PDFs have been selected, confirm with **[Open]**.

The [**More**>>] button offers several options:

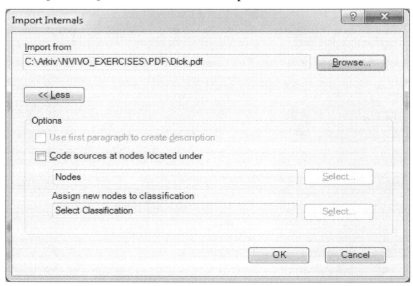

Use first paragraph to create descriptions: Not for PDF-items.

Code sources at new Nodes located under: Each Source Item will be coded at a Node (typically a Case Node) with the same name as the imported PDF file and located in a folder or under a parent Node that has been selected. Also you must assign the Nodes to a Classification when importing (see Chapter 11, Classifications).

7   Confirm the import with [**OK**].

When only one PDF has been selected the **PDF Properties** dialog box appears:

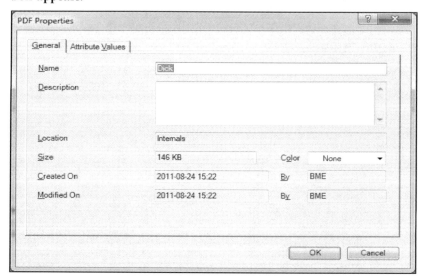

This dialog box will make it possible to modify the name of the PDF item and optionally add a description.

8   Confirm with [**OK**].

Opening a PDF Item

1   Click on [**Sources**] in Area 1.
2   Select the **Internals** folder in Area 2 or its subfolder.
3   Select the PDF in Area 3 that you want to open.
4   Go to **Home | Item | Open → Open PDF**
    or right-click and select **Open PDF...**
    or double-click on the PDF in Area 3
    or [**Ctrl**] + [**Shift**] + [**O**].

Sticky Notes in PDFs are very useful. You can create those with Acrobat Pro but also with recent versions of EndNote. Unfortunately NVivo cannot open these Notes. NVivo applies instead its link-tools, as a standard for all types of source items. Annotations serve the same purpose as the Sticky Notes.

Please note, that you can only open one PDF at a time, but several PDFs can stay open simultaneously.

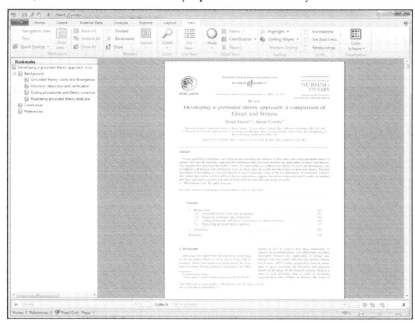

In this view you can code, link (See Also links, Annotations, Memo links) and search and query as with any other Source Item.

Selection Tools for PDF Items

There are two different selection tools for PDFs, Text or Region. Text Selection Mode is used for selecting any text in the PDF and is made as for any other selections within a Source Item. Selection Mode Text is default and is always active each time you open a PDF item.

Scanned text documents will not inherently allow selectable text; ensure you use software like Adobe Acrobat to recognize scanned text (OCR, Optical Character Recognition) so text selection is possible in NVivo.

Region Selection Mode is used when you need to select an image, a table or any graph that is in the PDF document. When you need to select an image, a table or any graph:

1. Open a PDF Source Item.
2. Go to **Home | Editing | PDF Selection → Region**
or point at the PDF right-click and select **Selection Mode → Region**.
3. With the mouse-pointer (which is now a cross) you define two diagonal corners of any rectangular area. Any text within such area will be interpreted as image not text.

To return to Selection Mode Text:

1. Go to **Home | Editing | PDF Selection → Region**
or point at the PDF, right-click and select **Selection Mode → Text**.

Selections made can now be used when coding and linking. Only hyperlinks cannot be created in a PDF Source Item. See page 174 on how a Node that codes a PDF item is shown.

Exporting a PDF Item

Like most NVivo items, PDF sources can also be exported:

1. Click on **[Sources]** in Area 1.
2. Select the **Internals** folder in Area 2 or its subfolder.
3. Select the PDF or PDFs in Area 3 that you want to export.
4. Go to **External Data | Export | Export → Export PDF...**
or right-click and select **Export | Export PDF...**
or **[Ctrl] + [Shift] + [E]**.

Tip: Working with PDF text documents. NVivo's functionality to work with PDF text documents can be a dream come true for researchers working on literature reviews. While many academic articles can be downloaded as functional PDF text documents, book chapters or other types of print material must often be scanned by researchers themselves. We recommend Adobe Acrobat Pro or ABBYY FineReader as software that will take scanned documents and recognize their text (a process called OCR, Optical Character Recognition).

The **Export Options** dialog box appears:

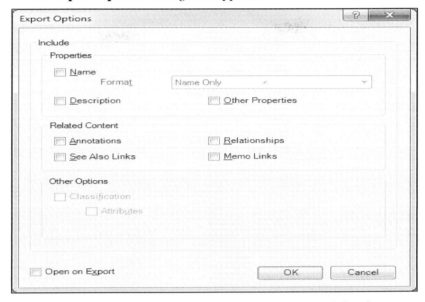

5 Select the options that you want. Confirm with **[OK]**.
6 Decide file name, file location, and file type. Possible file types are: .PDF and .HTML. The PDF file format is only available when none of the above options have been selected. Finally confirm with **[Save]**.

Please note, that any coding made on such items cannot be transferred when the Source Item is exported.

Tip: Using Word documents instead of PDFs. In our experience it is easier to work with Word files (.doc or .docx) than working with PDFs in NVivo. While it is not always possible to save your PDF files as Word documents, recent versions of Adobe Acrobat (X or XI) allow for PDF files to easily be saved as fully formatted Word files. Furthermore, Microsoft Word 2013 will allow for PDF files to be opened and saved as fully formatted Word files. *Also when you scan documents solely with the purpose of importing them to NVivo then create them as Word documents, which is the preferred file type.*

7. HANDLING AUDIO- AND VIDEO-SOURCES

So far, we have mainly focused on text data, but NVivo has a variety of useful functions for researchers interested in working with audio and video data. NVivo provides two primary functions for handling audio and video source data. First, audio and video data can be imported into NVivo as a data source, which can be organized, coded and queried similar to text source data. But second, and perhaps more importantly for some researchers, NVivo contains a full functioning transcription utility for importing, creating, and exporting text transcripts. Instead of outsourcing transcription to third-party vendors or spending funds on specialized transcription software, NVivo gives researchers a very useful option for transcribing their own audio and video files within the software.

NVivo 10 can import the following audio formats: .MP3, .M4A, .WAV, and .WMA and the following video formats: .MPG, .MPEG, .MPE, .MP4, .MOV, .QT, .3GP, .MTS, and .M2TS. Several of these media formats are new to NVivo 10 to allow users to import more media content form their smart phones. Media files less than 40 MB can be imported and embedded in you NVivo project.

Files larger than 40 MB must be stored as external files. Importantly, external files can be handled the same way as an audio or video embedded item. NVivo contains an on-board audio and video player for external files, so even though a large video file may not be embedded in your project, you can still view, transcribe, code, and query the file using the NVivo player. But remember, if you open your NVivo project on another computer the external file references will no longer work, unless you assemble copies of those files in identically named file folders on the new computer you are using.

Even the not embedded media items are located under the **Internals** folder or its subfolders as they are managed in all respects as if they were embedded.

The threshold value for audio and video files that can be stored as external files can be reduced for all new projects with **File → Options...**, select the **Audio/Video** tab, section **Default for new projects** (see page 44). To adjust values for the current project, use **File → Info → Project Properties...**, select the **Audio/Video** tab, section **Settings** (see page 57). To adjust for the current Audio/Video item, go to **Audio/Video Properties** dialog box, the **Audio/Video** tab (see page 94).

When you need to view all items that are not embedded go to [**Folders**] in Area 1, select **Search Folders** and subfolder **All Sources Not Embedded** in Area 2.

Importing Media Files

Importing media files follows a similar, simple protocol as importing text files or PDFs:
1. Go to **External Data | Import | Audios/Videos**
   Default folder is **Internals**.
   Go to 5.

alternatively
1. Click **[Sources]** in Area 1.
2. Select the **Internals** folder in Area 2 or its subfolder.
3. Go to **External Data | Import | Audios/Videos**.
   Go to 5.

alternatively
3. Click on any empty space in Area 3.
4. Right-click and select **Import Internals → Import Audios.../Import Videos...**
   or **[Ctrl]** + **[Shift]** + **[I]**.

alternatively
3. Drag and drop your file's icon from an outside folder into Area 3.
   Go to 5.

The **Import Internals** dialog box appears:

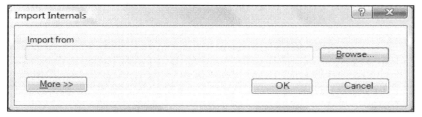

5. The **[Browse...]** button gives access to a file browser and you can select one or several media files for a batch import.
6. When the file or files have been chosen, confirm with **[Open]**.

The [**More** >>] button lets you select more options:

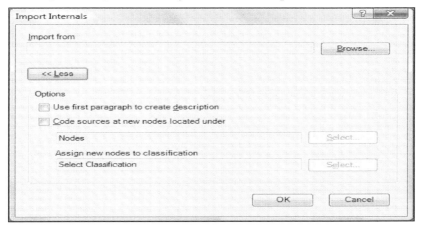

Use first paragraph to create descriptions: Not applicable for the import of media files.

Code sources at new Nodes located under. Each Source Item will be coded at a Node (typically a Case Node) with the same name as the imported file and located under in a folder or under a parent Node that has been selected. Also you must assign the Nodes to a Classification when importing (see Chapter 11, Classifications).

7   Confirm the import with [**OK**].

When only one media file is imported the **Audio Properties/Video Properties** dialog box is shown:

This dialog box will make it possible to modify the name of the item and optionally add a description.

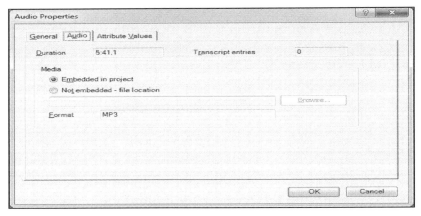

When the **Audio/Video** tab has been chosen you can let the audio file be stored as an external file even if the size is below the limit for embedding. After an audio- or video file has been imported you can change the properties from embedded item to external storage and vice versa by using **Audio Properties/Video Properties**. An embedded item cannot exceed 40 MB.

8   Confirm with **[OK]**.

At times you may need to move an external media file. When an external file has been moved the media item must be updated through NVivo. Go to **Home | Item | Properties → Update Media file Location** or rightclick and select **Update File Location** and select the external file's new location. From there, NVivo will find the correct media file.

Creating a New Media Item

Instead of importing an audio or video item, a new media item can also be created:

1   Go to **Create | Sources | Audio/Video**.
Default folder is **Internals**.
Go to 5.

alternatively

1   Click **[Sources]** in Area 1.
2   Select the **Internals** folder in Area 2 or its subfolder.
3   Go to **Create | Sources | Audio/Video**.
Go to 5.

alternatively

3   Click on any empty space in Area 3.
4   Right-click and select **New Internal → New Audio.../New Video...**
or **[Ctrl] + [Shift] + [N]**.

The **New Audio/New Video** dialog box appears:

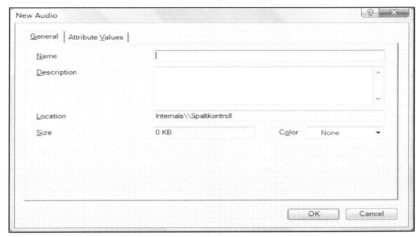

5   Type name (compulsory) and a description (optional), then [**OK**].

When you create a new media item it initially has no media file or transcript. Instead these pieces of information can be imported separately. From the open media item, click the **Edit Mode** link and go to **Media | Import | Media Content** or **Media | Import | Transcript Rows** (see page 100). From here, select the required contents.

Here is a typical list view in Area 3 of some audio items:

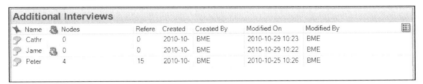

Opening a Media Item

Now that you've created and imported some media items you'll want to open them to access their data. Crucially, when handling media items you will have access to the media ribbon, one of NVivo's context-dependent ribbons:

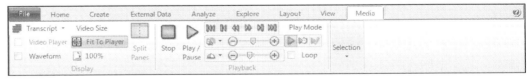

1   Click [**Sources**] in Area 1.

2 Select the **Internals** folder in Area 2 or its subfolder.
3 Select the media i
4 tem in Area 3 that you want to open.
5 Go to **Home | Item | Open**
   or right-click and select **Open Audio/Video...**
   or double-click on the media item in Area 3
   or **[Ctrl] + [Shift] + [O]**.

Please note, NVivo only allows you to open one media item at a time, but several items can stay open simultaneously.

An open audio item may appear like this:

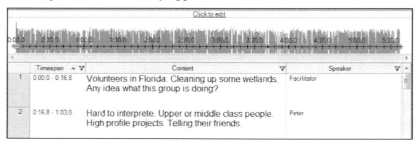

Provided a soundcard and speakers are connected to the computer you can now play and analyze the audio item.

Creating Custom Transcript Fields

When you import transcripts, see page 100, then the transcript file may have defined the Custom Transcript Fields. Otherwise, if you have defined those fields with **Application Options**, the **Audio/Video** tab, page 44, then all new projects will have them. If you need to define or edit the Transcript Fields for the current project, then use **Project Properties**, the **Audio/Video** tab, see page 57.

A practical arrangement is to use **Media | Display | Split Panes** which separates the default fields from the Custom Transcript Fields. This makes it easier to adjust the column widths.

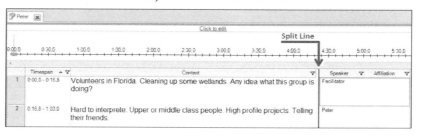

Play Modes

NVivo offers three Play Modes for working with media items, with each having a special function relating to transcription. *Normal Mode* simply plays your media item; *Synchronize Mode* plays your media item while scrolling through the corresponding rows of your transcript; and *Transcribe Mode* creates a new time interval each time you play your media item, and ends that interval when you stop.

Go to **Media | Playback | Playmode** to view or change playmode options.

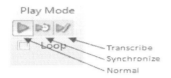

Playmode Normal
When a media item is opened the play mode is always *Normal*.
1. Go to **Media | Playback | Play/Pause** or [**F4**].

Only the selected section will be played if there is a selection along the timeline. The selection disappears when you click outside the selection.
1. Go to **Media | Playback | Stop** or [**F8**].

Rewind, Fast Forward etc.
1. Go to **Media | Playback | Go to Start**.
2. Go to **Media | Playback | Rewind**.
3. Go to **Media | Playback | Fast Forward**.
4. Go to **Media | Playback | Go to End**.
5. Go to **Media | Playback | Skip Back**. or [**F9**]
6. Go to **Media | Playback | Skip Forward**. or [**F10**]

The *Skip* interval is determined by the setting under **File → Options**, the **Audio/Video** tab (see page 44).

Volume and Speed
1. Go to **Media | Playback | Volume**. This slider also allows mute.
2. Go to **Media | Playback | Play Speed**. There are fixed positions and continuous slider.

Play Mode Synchronized
You can play any media item synchronized so the transcription text row is highlighted and always visible (by automatic scrolling).
1. Go to **Media | Playback | Play Mode → Synchronize**.
2. Play.

Play Mode Transcribe
You can link audio timeline intervals with rows of text (e.g., written comments, direct transcripts or translations). In NVivo, the practice of linking time segments of audio or video with rows of text is called *transcription*. While *transcription* can be used to create verbatim transcripts of your audio files, some researchers find it faster to write short-hand transcripts.

Transcription requires several steps. First, you need to define an interval that will correspond with the row of the text. Next, the audio timeline interval and the text row need to be linked. From there, you have a transcript ready to code and link.
1. Go to **Media | Playback | Play Mode → Transcribe**.
2. Play.

> **Tip:** We recommend hiding the waveform to make it easier to view a selection and other markings along the timeline. Go to **Media | Display | Waveform**, which is a toggling function. Each media item retains its individual setting during the ongoing session.
>
> You may set a default for viewing the waveform by going to **File → Options**, the **Display** tab and deselect *Waveform*.

Timeline intervals can be defined in a number of ways, such as by selecting portions of the timeline with your mouse when audio is paused, or by using keyboard commands to mark the start and end of an interval while audio is playing (our preferred method!). See page 100.

Selecting a Time Interval in Play Mode Normal
NVivo acts like a simple audio file player when in Normal Play Mode. There are *two ways* to select a time interval that *can be used for* coding or for creating a transcription row. Please note that these methods will work in *any* P*lay Mode:*
1. Use the left mouse button to define the start of an interval, then hold the button, drag along the timeline, and release the button at the end of the interval.

alternatively
1. Play the media item, possibly at low speed, see above.
2. Determine the start of an interval by going to
   Media | Selection | Start Selection
   or **[F11]**.
3. Determine the end of an interval by going to
   Media | Selection | Stop Selection
   or **[F12]**.

The result is a selection (a blue frame) along the timeline. Now you can code or link from this selection. To proceed with creating the next selection you need to click outside the previous selection. Retaining the current selection will limit the play interval to that selection.

Creating a Transcript Row from a Time Interval in Normal Play Mode
Once you have selected a time interval, there are several methods to create a new row. Please note that these methods will work in *any* Play Mode:
1. Make a selection along the timeline.
2. Go to **Layout | Rows and Columns | Insert → Insert Row**
   or right-click and select **Insert Row**
   or **[Ctrl] + [Ins]**.

The result is a transcript row corresponding to the selected time interval called *Timespan*, with the textbox in the column called *Content*:

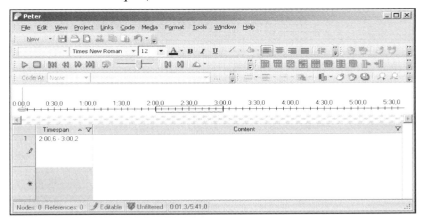

Should you need to adjust a timespan you can do as follows:
1. Select a transcript row by clicking in the row item number (the leftmost column). The corresponding timespan along the timeline is then marked with a purple guiding line.
2. Make a new modified selection along the timeline.
3. Go to **Media | Selection | Assign Timespan to Rows**
   or right-click and select **Assign Timespan to Rows**.

As an alternative you can also modify the timespan directly in the transcript row by typing a new start time and a new stop time. From there you can then make a new selection along the timeline
1. Select a transcript row by clicking in the row item number (the leftmost column).
2. Go to **Media | Selection | Select Media from Transcript**.

99

Creating a Transcript with Transcribe Mode

As experienced NVivo users and trainers, we believe that Transcribe Mode is the best method for researchers who are using media files to create verbatim transcripts, real-time summaries, or notes on extra-linguistic cues or vocal intonation. Transcribe Mode allows you, using basic keyboard shortcuts, to quickly and easily generate text to accompany your media data.

1. Go to **Media | Playback | Play Mode → Transcribe**.
2. Play and determine start of an interval by going to **Media | Playback | Start/Pause** or **[F4]**.
3. Determine end of an interval by going to **Media | Playback | Stop** or **[F8]**.

While transcribing you can pause the audio if you need time to finish writing. We recommend going to **File → Options**, the **Audio/Video** tab and turning on the setting *Skip back on play in transcribe mode*. This will automatically rewind an interval you have created back to its beginning after you pause your transcription.

At any time you can also create the beginning of a new interval with **[F11]** and then end it with **[F12]**. You will also get a new transcript row but you need to pause with a separate command.

Merging Transcript Rows

Sometimes there is a need of cleaning up or reducing the number of transcript rows by merging several rows:

1. Open a media item in edit mode.
2. Select two or more transcript rows by holding down the **[Ctrl]** key and left-clicking in the item number column of the transcript rows.
3. Go to **Layout | Rows & Columns | Merge Rows**.

The merged row now covers the timespan from the first to the last selected timeslots.

Importing Transcripts

In the event your transcripts are existing text files on your computer (perhaps you are fortunate enough to be using a transcription service for your project), it is possible to import text material as a transcript for its original audio file. NVivo allows you to correspond your transcript text with the audio file by using either Timestamps, Paragraphs, or Tables. The file format of your imported text file needs to be .DOC, .DOCX, .RTF, or .TXT.

The *Timestamp* Style format:

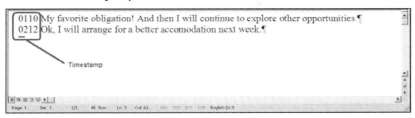

The *Paragraph* Style format:

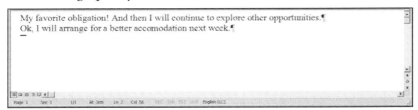

The *Table* Style format:

Timespan	Content	Speaker
0110	My favorite obligation! And then I will continue to explore other opportunities.	Ruth
0212	Ok, I will arrange for a better accomodation next week.	Edgar

To import a transcript file:
1. Open the media item in edit mode.
2. Go to **Media | Import | Transcript Rows**.

The **Import Transcript Entries** dialog box appears:

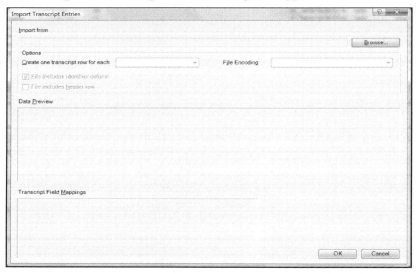

3   The **[Browse...]** button gives access to a file browser and you can select the file you want to import.
4   Once a file is selected at *Options, Create one transcript row for each* you need to select an alternative that corresponds to the appropriate style format.

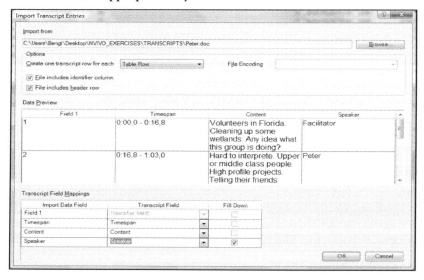

5   When Data Preview displays a correct image of the transcript then you need to set the Transcript Field Mappings so that imported data are mapped to the proper columns in the media item.
6   Confirm with **[OK]**.

Note, when more columns are included in the imported file these columns are also created in NVivo:

#	Timespan	Content	Speaker
1	0:00,0 - 0:16,8	Volunteers in Florida. Cleaning up some wetlands. Any idea what this group is doing?	Facilitator
2	0:16,8 - 1:03,0	Hard to interprete. Upper or middle class people. High profile projects. Telling their friends.	Peter
3	1:03,0 - 1:23,5	Con is working with teenagers on a Youth center. Helping them with career choises.	Facilitator
4	1:23,5 - 2:03,0	Working with problematic teenagers is not very fun, but he thinks he can make a contribution to his society.	Peter

Transcript Display Options

Transcript rows can also be hidden:
1. Go to **Media | Display | Transcript → Hide**.

This is a toggling function.

You may also hide/unhide the video player in a video item:
1. Go to **Media | Display | Video Player**.

This is a toggling function.

For video items you may also want to display the transcript rows below the media timeline and the video player:
1. Open a video item.
2. Go to **Media | Display | Transcript → Bottom**.

> **Tip:** Are you already working with a transcription service? Most transcriptionists can provide timestamped transcripts at no additional charge. You can easily import your transcripts to correspond with your project media items if you work with your transcriptionist to ensure they are formatting the timestamps for NVivo.

Coding a Media Item

With your newly created time intervals or transcript rows, you may want to begin coding data to correspond with project Nodes (Chapter 10, Introducing Nodes). Coding a media item can be done in two ways:
1. Coding the transcript row or words in the transcript text
2. Coding a timeslot along the timeline

These coding principles are the same for media items as for any text material: select a text or an interval that to code and then select the Node or Nodes at which you will be coding.

If you want to code a whole transcript row, select the row by clicking the item number column, then right-click and select the Node or Nodes that you want to code at.

If you want to code a certain timeslot along the timeline, make a selection and then select the Node or Nodes, see Chapter 10, Introducing Nodes and Chapter 12, Coding.

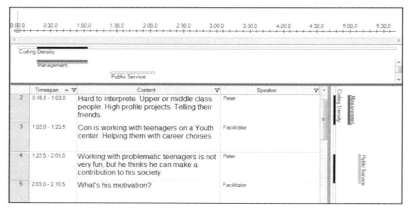

Shadow Coding

Shadow Coding is a special feature related to coding of media items. Shadow coding means that when a text or a row in a transcript has been coded the corresponding interval of the timeline displays the coding faintly, like a shadow. Shadow coding can only be shown with coding stripes turned on (see page 167). Coding stripes are filled colored lines and shadow coding stripes are the same colour, but lighter – hence the name shadow coding.

The media item above is coded at the Nodes *Management* and *Public Service*. Both the transcript row and the timeslot are coded at the Node *Management*. Therefore the item has 'double' coding stripes. The Node *Public Service* is only coded at the transcript row. Shadow coding has no use other than being a visual aid when studying coding stripes. Shadow coding can be switched on and off with **View | Coding | Coding Stripes → Shadow Coding**.

Working with the Timeline

Sometimes there is a need of selecting a timespan from an existing transcript:
1. Open a media item with transcript rows.
2. Select a transcript row.
3. Go to **Media | Select | Select Media from Transcript**.

Now there is an exact selection and you can play, code or link from this selection.

Playing an interval from a transcript row:
1. Open a media item with transcript rows.
2. Select a transcript row.
3. Go to **Media | Selection | Play Transcript Media**.

Only the selected interval will be played.

When you open a Node that codes both a row and a timeslot then click on a coding stripe, open the **Audio** tab and it looks like this. Playing from here only plays the coded timeslot(s).

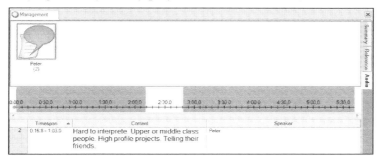

About autocoding of transcripts, see page 159.

Linking from a Media Item

An audio item can be linked (Memo Links, See Also Links and Annotations) in the same manner as any other NVivo item. However, hyperlinks cannot be created from an audio item. Links can be created from a selected timespan or from the transcript.

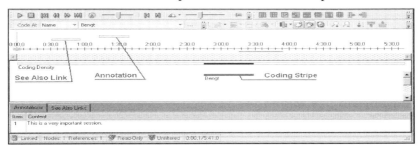

A Memo Link is shown in the list view. A See Also Link or an Annotation that refers to a timespan are shown above the timeline as a filled pink line and a filled blue line respectively. Coding stripes are shown below the timeline, see Chapter 9, Memos, Links, and Annotations.

Exporting a Media Item

Like any Source Item in NVivo, media items can be exported:
1. Click **[Sources]** in Area 1.
2. Select the **Internals** folder in Area 2 or its subfolder.
3. Select the media item or items in Area 3 that you want to export.
4. Go to **External Data | Export | Export →
   Export Audio(Video)/Transcript...**
   or right-click and select **Export →
   Export Audio(Video)/Transcript...**
   or **[Ctrl] + [Shift] + [E]**.

The **Export Options** dialog box appears:

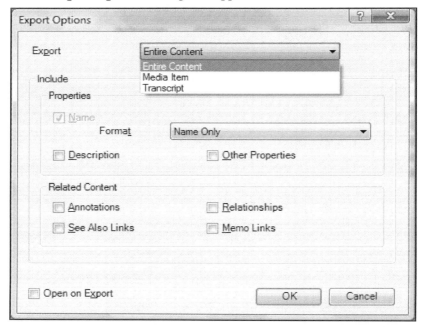

5   Select applicable options for the export of the media file, the transcript, or both. Confirm with **[OK]**.
6   Decide the file path, file type (*.HTM, *.HTML) and filename, confirm with **[Save]**.

When you select *Entire Content* the result is a web page and the media file and other supporting files are stored in a folder called 'Filename_files'. If you also select the option *Open on Export* then the web browser opens and the result is shown instantly.

- ♦ -

A media item can also be printed with the normal command **File → Print → Print** or **[Ctrl]** + **[P]**. A transcript and its coding stripes can similarly be printed.

8. HANDLING PICTURE SOURCES

In the same way that NVivo associates media sources with timespans which correspond to text (e.g., transcription rows), handling pictures in NVivo is about defining a Region of the picture which then can be associated with a written note, called a Picture Log. Both a Region and a Picture Log can be coded and linked. NVivo 10 can import the following picture formats: .BMP, .GIF, .JPG, .JPEG, .TIF and .TIFF.

Importing Picture Files

NVivo can easily import a number of the most common image types. Plenty of free online image converter websites exist in the event you find you possess an image file that is a different format than NVivo accepts:

1. Go to **External Data | Import | Import Pictures**.
   Default folder is **Internals**.
   Go to 5.

alternatively

1. Click **[Sources]** in Area 1.
2. Select the **Internals** folder in Area 2 or its subfolder.
3. Go to **External Data | Import | Import Pictures**.
   Go to 5.

alternatively

3. Click on any empty space in Area 3.
4. Right-click and select **Import Internals → Import Pictures...**
   or **[Ctrl] + [Shift] + [I]**.

alternatively

3. Drag and drop your file's icons from an outside folder into Area 3.
   Go to 5.

The **Import Internals** dialog box appears:

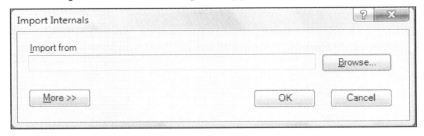

5. The **[Browse...]** button gives access to a file browser and you can select one or several picture files for import.
6. When the picture files have been selected, confirm with **[Open]**.

The [**More** >>] button offers several options:

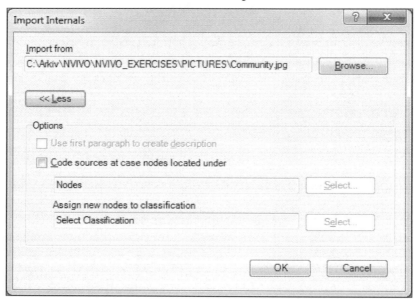

Use first paragraph to create descriptions. Not applicable for picture files.

Code sources at new Nodes located under. Each Source Item will be coded at a Node (typically a Case Node) with the same name as the imported file and located in a folder or under a parent Node that has been selected. Also you must assign the Nodes to a Classification when importing (see Chapter 11, Classifications).

7   Confirm the import with [**OK**].

When only *one* picture file has been imported the **Picture Properties** dialog box is shown:

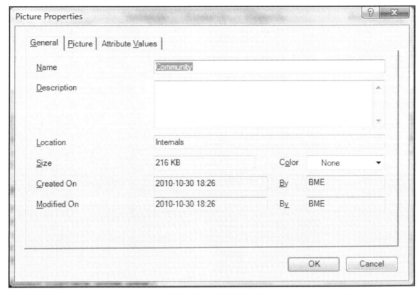

This dialog box makes it possible to modify the name of the item and optionally add a description. The **Picture** tab gives access to details and data from the imported picture:

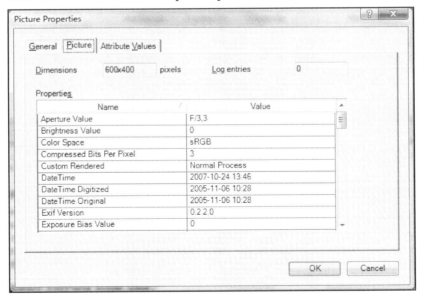

8   Confirm with [**OK**].

Here is a typical list view in Area 3 of some picture items:

Opening a Picture Item

1 Click **[Sources]** in Area 1.
2 Select the **Internals** folder in Area 2 or its subfolder.
3 Select the picture item in Area 3 that you want to open.
4 Go to **Home | Item | Open**
or right-click and select **Open Picture...**
or double-click the picture item in Area 3
or **[Ctrl] + [Shift] + [O]**.

The **Picture** ribbon now opens. Remember that NVivo can only open one picture item at a time, but several picture items can stay open simultaneously.

An open picture item can look like this:

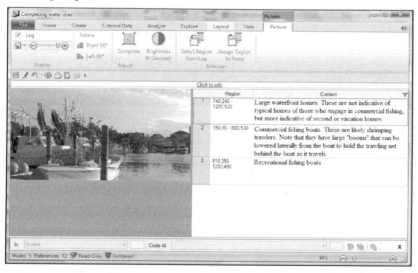

The handling of pictures is about defining a Region of the picture which then can be associated with a written note, a Picture Log. Both a Region and a Picture Log can be coded and linked.

Selecting a Region and Creating a Picture Log

1. Select a corner of the Region with the left mouse button, then drag the mouse pointer to the opposite corner and release the button.
2. Go to **Layout | Rows & Columns | Insert → Row** or **[Ctrl] + [Ins]**.

The result can appear like this and the Picture Log can be typed in the cell below the column head Content:

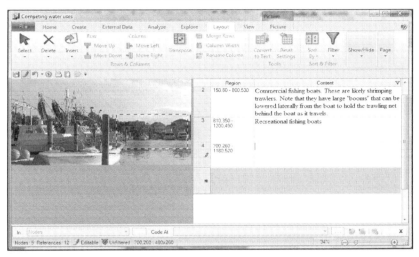

Sometimes you may need to redefine a Region and a Picture Log:
1. Select the row of the Picture Log that you wish to redefine. When selecting a Row the corresponding Region is highlighted.
2. Select a new Region (redefine a highlighted area).
3. Go to **Picture | Selection | Assign Region to Rows**.

In this way you adjust both a Region and a Row of the Picture Log.

As an alternative you can use a Row from which you can select a new Region:
1. Select a Row of the Picture Log. Corresponding region will be highlighted.
2. Go to **Picture | Selection | Select Region from Log**.

You can also hide the Picture Log:
1 Go to **Picture | Display | Log**.
This is a toggling function and a Picture item with a hidden Picture Log appears like this:

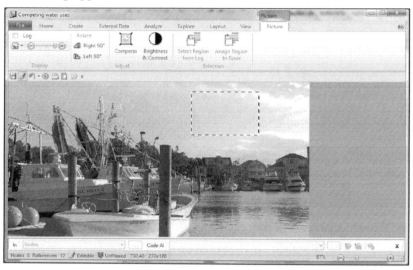

Editing Pictures

NVivo offers some basic functions for easy editing of picture items. The following functions are menu options after a picture item has been opened:

Picture | Adjust | Rotate → Right 90°
Picture | Adjust | Rotate → Left 90°
Picture | Adjust | Compress
Picture | Adjust | Brightness & Contrast

Coding a Picture Item

You can code a Picture Log, a selected text element or a Region of a picture item. The act of coding is in principle the same way you would code other elements of your NVivo project. In short you select data to be coded and then you select the Node or Nodes that the data will be coded at, see Chapter 10, Introducing Nodes and Chapter 12, Coding.

If you need to code a row of the picture log, select the row by clicking the leftmost column of the row, right-clicking and selecting a Node or Nodes.

If instead (or in addition) you need to code a Region, select the Region and select a Node or Nodes as usual.

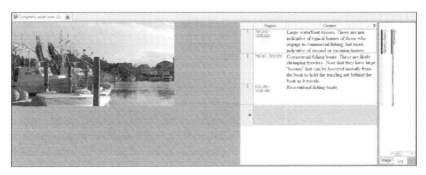

Coding stripes for a picture item are always shown in a window to the left and in a position leveled with the region. Coding stripes from a coded region are colored and filled while coding stripes from a coded picture log are lighter colored. (Same look as Shadow coding stripes.)

The above example shows a picture item that has been coded at the Node *The Star*. Both the region and the picture log have been directly coded and therefore 'double' coding stripes are shown.

We would also like to show when the Node *Management* has been opened. After having clicked the **Picture** tab to the right you will see both the coded region of the picture and the corresponding Picture Log.

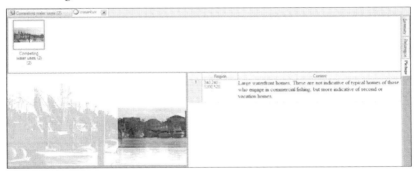

Linking from a Picture Item

A picture item can be linked (Memo Links, See Also Links and Annotations) in the same manner as any other NVivo item. However, hyperlinks cannot be created from a picture item. Links can be created from a selected region or from the Picture log. A Memo Link is not shown elsewhere than in the list view. See Also Links or Annotations are shown as pink and blue frames respectively:

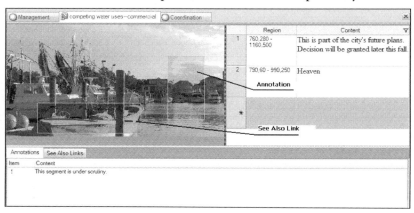

See also Chapter 9, Memos, Links, and Annotations.

Exporting a Picture Item

1. Click [**Sources**] in Area 1.
2. Select the **Internals** folder in Area 2 or its subfolder.
3. Select the picture item or items in Area 3 that you want to export.
4. Go to **External Data | Export | Export → Export Picture/Log**
   or right-click and select **Export → Picture/Log...**
   or [**Ctrl**] + [**Shift**] + [**E**].

The **Export Options** dialog box appears:

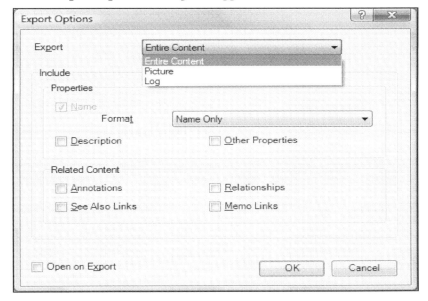

5. Select applicable options, which allows for the export of . the picture file, the picture log or both. Confirm with [**OK**].
6. Decide file name, file location, and file type, confirm with [**Save**].

When you select *Entire Content* the result is a web page with the picture file in a folder called 'Filename_files'. Your web browser will open and the result is instantly shown if you also check the option *Open on Export*.

- ♦ -

A picture item can also be printed with the normal command **File → Print → Print** or [**Ctrl**] + [**P**]. The picture, and optionally the transcript and the coding stripes can be printed.

9. MEMOS, LINKS, AND ANNOTATIONS

Memos, Memo Links, See Also Links, Hyperlinks and Annotations are NVivo tools that allow you to create connections and track your ideas across your data. While similar in function, each of these four tools operates differently, with Memos and Memo Links being closely related.

Exploring Links in the List View

Memo Links, See Also Links, and Annotations (but not Hyperlinks) can be opened and viewed in List View in Area 3 like any other Project Item.
1 Click [**Folders**] in Area 1.
2 Select any of the following folders in Area 2:
Memo Links
See Also Links
Annotations

Next you will see the selected list of links as items in Area 3.

Right-clicking a **Memo Link** item in Area 3 will open a menu with the options: Open Linked Item, Open Linked Memo or Delete Memo Link. Exporting and printing the whole list of items are also available options.

Double-clicking a **See Also Link** in Area 3 opens the **See Also Link Properties** dialog box. Right-clicking on such item will open a menu with the options: Open From Item, Open To Item, Edit See Also Link or Delete See Also Link. Exporting and printing the whole list of items are also available options.

Double-clicking an **Annotation** item in Area 3 opens the source and its Annotation in Area 4. Right-clicking on an Annotation will open a menu with the options: Open Source and Delete Annotation. Exporting and printing the whole list of items are also available options.

Memos

Memos are a type of source that allows you to record research insights in a source document that can be linked to another item in your project. Any Source or Node can have one Memo linked to it, called a Memo Link. For example, Memos can be notes, instructions or field notes that have been created outside NVivo. A memo cannot be linked to another memo with a Memo Link.

Importing a Memo

As with other Project Items you can import them or create them with NVivo. The following file formats can be imported as memos: .DOC, .DOCX, .RTF, and .TXT.

    1   Go to **External Data | Import | Memos**.
        Default folder is **Memos**.
        Go to 5.

alternatively

    1   Click **[Sources]** in Area 1.
    2   Select the **Memos** folder in Area 2 or its subfolder.
    3   Go to **External Data | Import | Memos**.
        Go to 5.

alternatively

    3   Click on any empty space in Area 3.
    4   Right-click and select **Import Memos...**
        or **[Ctrl]** + **[Shift]** + **[I]**.

alternatively

    3   Drag and drop your file's icon from an outside folder into Area 3.
        Go to 5.

The **Import Memos** dialog box appears:

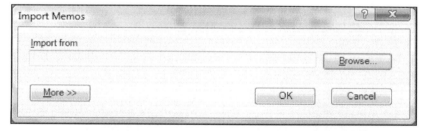

    5   The **[Browse...]** button gives access to a file browser and you can select one or several documents for a batch import.
    6   When the documents have been selected, confirm with **[Open]**.

The [**More** >>] button gives access to several options:

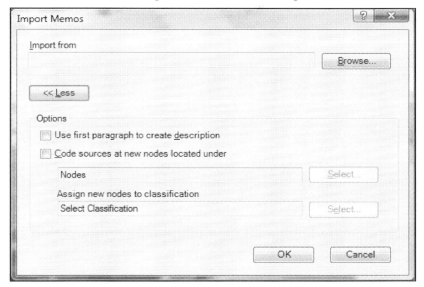

Use first paragraph to create description: NVivo copies the first paragraph of the document and pastes it into the description text box.

Code sources at new Nodes located under: Each Source Item will be coded at a Node (typically a Case Node) with the same name as the imported file and located in a folder or under a parent Node that has been selected. Also you must assign the Nodes to a Classification when importing (see Chapter 11, Classifications).

    7   Confirm the import with [**OK**].

Creating a Memo

    1   Go to **Create | Sources | Memo**.
           Default folder is **Memos**.
           Go to 5.

alternatively

    1   Click [**Sources**] in Area 1.
    2   Select the **Memos** folder in Area 2 or its subfolder.
    3   Go to **Create | Sources | Memo**.
           Go to 5.

alternatively

    3   Click on any empty space in Area 3.
    4   Right-click and select **New Memo...**
        or [**Ctrl**] + [**Shift**] + [**N**].

The **New Memo** dialog box appears:

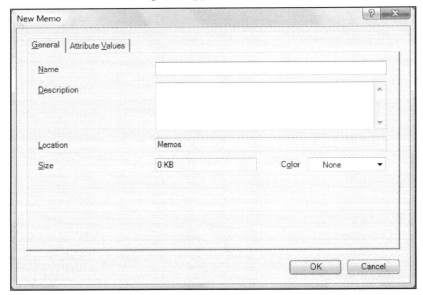

5   Type a name (compulsory) and a description (optional), then [**OK**].

Here is a typical list view in Area 3 of some memos:

Opening a Memo
1   Click [**Sources**] in Area 1.
2   Select the **Memos** folder in Area 2 or its subfolder.
3   Select a memo in Area 3 that you want to open.
4   Go to **Home | Item | Open**
    or right-click and select **Open Memo...**
    or double-click on the memo in Area 3
    or [**Ctrl**] + [**Shift**] + [**O**].

Please note, you can only open one memo at a time, but several memos can stay open.

Creating a Memo Link

Memo Links truly distinguish Memos from other types of NVivo sources. Memo Links are an optional component of Memos.

1. In the list view, Area 3, select the item from which you want to create a Memo Link. You cannot create a Memo Link to a memo that is already linked.
2. Go to **Analyze | Links → Memo Link → Link to Existing Memo...**
   or right-click and select **Memo Link → Link to Existing Memo...**

The **Select Project Item** dialog box is shown. Only unlinked memos can be selected, linked memos are dimmed.

3. Select the memo that you want to link to and confirm with **[OK]**.

The Memo Link is shown in the list view in Area 3 with one icon for the memo and one icon for the linked item.

Creating a Memo Link and a New Memo Simultaneously

NVivo makes it easy to create a Memo Link and a new Memo simultaneously:

1. In the list view, Area 3, select the item from which you want to create a Memo Link and a new Memo.
2. Go to **Analyze | Links → Memo Link → Link to New Memo...**
   or right-click and select **Memo Link → Link to New Memo...**
   or **[Ctrl] + [Shift] + [K]**.

The **New Memo** dialog box is shown and you continue according to page 120.

Opening a Linked Memo
A Memo can be opened as outlined above, but a linked Memo can also be in the event where a Memo Link is in place.
1. In the list view, Area 3, select the item from which you want to open a Linked Memo.
2. Go to **Analyze | Links → Memo Link → Open Linked Memo**
   or right-click and select **Memo Link → Open Linked Memo** or **[Ctrl] + [Shift] + [M]**.

Deleting a Memo Link
1. In the list view, Area 3, select the item from which you want to delete a Memo Link.
2. Go to **Analyze | Links → Memo Link → Delete Memo Link**
   or right-click and select **Memo Link → Delete Memo Link**.

The **Delete Confirmation** dialog box appears:

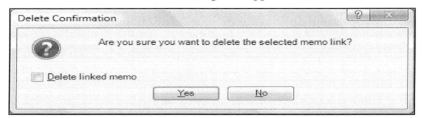

3. If you select *Delete linked memo* then also the Memo will be deleted, otherwise only the Memo Link will be deleted. Confirm with **[Yes]**.

See Also Links

See Also Links literally create connection points between different items from your NVivo project. See Also Links are links from a selection (text, picture region, or audio segment) in an item to another item or a certain selection from another item. Multiple See Also Links can be linked to the same item, unlike Memo Links which are a one-to-one relationship between a Memo and its attendant NVivo item.

Creating a See Also Link to Another Item
1. Open the item from which you want to create a See Also Link.
2. Select the section (text, picture) from which you want to create a See Also Link.
3. Go to **Analyze | Links → See Also Link → New See Also Link...**
   or right-click and select **Links → See Also Link → New See Also Link...**

The **New See Also Link** dialog box appears:

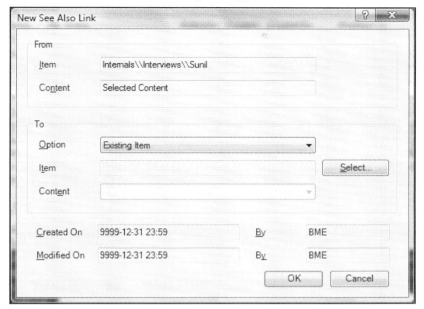

Under the **Option** drop down list, you can select what type of Project Item you will link. An item will be created if you select an option starting with New then. If you select the option Existing Item you go to the [**Select...**] button and use the **Select Project Item** dialog box to select an item to link to. When the item has been selected the link goes to the entire target item.

4   Confirm with [**OK**].

For example, in this Source Item the See Also Link is indicated as a pink colored highlighting:

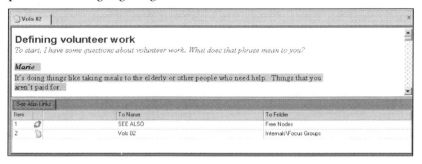

Creating a See Also Link to a Selection of Another Item
1. Open the target item that you want to link to.
2. Select the section (text, image) that you want to link to.
3. Copy with for example [**Ctrl**] + [**C**].
4. Open the item from which you want to create a See Also Link.
5. Select the area (text, image) that you want to link from.
6. Go to **Home | Clipboard | Paste → Paste As See Also Link**.

Opening a See Also Link
1. Position the cursor on the See Also Link or select the entire link.
2. Go to **Analyze | Links | Se Also Links → Open To Item** or right-click and select **Links → Open To Item**.

The target item will open and if you have used the option *Selected Content* the selected area will be shown highlighted otherwise not.

Hiding or Unhiding See Also Links
You can view all the See Also Links from a certain item in a window below the open item. The links are shown as a list of items. Clicking on an item opens the link. Right-clicking and selecting **Open To Item...** also opens the link.
1. Open the item that has one or several See Also Links.
2. Go to **View | Links | See Also Links**.

This is a toggling function for hiding or unhiding the See Also Link window.

Opening a Linked External Source
Provided the See Also Link leads to an external item, you are able to open that external source (file or web site) directly. You may wish to create links to external sources rather than creating hyperlinks as you may reduce unnecessary modifications to the external sources.

1. Position the cursor on the See Also Link or select the entire link.
2. Go to **Analyze | Links | See Also Links → Open Linked External File**
or right-click and select **Links → Open Linked External File**.

Deleting a See Also Link
1. Position the cursor on the See Also Link or select the entire link.
2. Go to **Analyze | Links | See Also Link → Delete See Also Link**
or right-click and select **Links → See Also Link → Delete See Also Link.**
3. Confirm with **[Yes]**.

Annotations

Annotations and See Also Links are similar but different. When you create an Annotation, the Annotations tab appears at the bottom of a Source or Node at a point of your choosing. An annotations could be a quick note, a reference or an idea. Unlikle Memos, which can only link to entire sources, Annotations link to specific segments of your data (e.g., text from a focus group transcript, or a segment of time from a video source). An annotation in NVivo shares similarities with a footnote in Word, especially because annotations are numbered within each Project Item.

Creating an Annotation
1. Open a Source Item or a Node.
2. Select the text or other section area that you want to link to an Annotation.
3. Go to **Analyze | Annotation → New Annotation...**
or right-click and select **Links → Annotation → New Annotation**.

A new window will open where the annotation can be typed. The linked area is then shown highlighted in blue.

Does it fit with your goals? Do you expect to have enough time to do what you want to do?
Yes that's really the goal. Spare time is a dream.

Annotations	
Item	Content
1	

Hiding or Unhiding Annotations
When an NVivo item contains Annotations, you have the option of toggling the Annotations tab on or off.
1. Open the Project Item that contains Annotations.
2. Go to **View | Links | Annotations**.

This is a toggling function and is valid separately for each item.

Deleting an Annotation
1. Position the cursor on the link to an Annotation.
2. Go to **Analyze | Annotation | Delete Annotation** or right-click and select **Links → Annotation → Delete Annotation**.
3. Confirm with [**Yes**].

Hyperlinks
NVivo can create links to external sources in two ways:
- Hyperlinks from a Source Item.
- External items (see page 68).

Creating Hyperlinks
1. Select a section (text or image) in a Source Item while in Edit mode.
2. Go to **Analyze | Links | Hyperlink → New Hyperlink...** or right-click and select **Links → Hyperlink → New Hyperlink...**

The **New Hyperlink** dialog box appears:

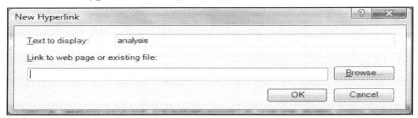

3. Paste a complete URL or use [**Browse...**] to find the target file in your computer or in your local network.
4. Confirm with [**OK**].

A Hyperlink is blue and underlined.

Opening a Hyperlink
The following three methods will open a Hyperlink:
1. Position the cursor on the link.
2. Go to **Analyze | Links | Hyperlink → Open Hyperlink**.

alternatively
1. Point at the link with the mouse pointer which will then become an arrow.
2. Right-click and select **Links → Hyperlink → Open Hyperlink**.

alternatively
1. Hold down the **[Ctrl]** key.
2. Click on the link.

This latter command will sometimes cause the external file (depending on the file type) to open as a minimized window. If this is the case then you can either repeat this command or click on the program button of the Windows toolbar so the window opens fully.

Deleting a Hyperlink
1. Position the cursor on the link in a Source Item while in Edit mode.
2. Go to **Analyze | Links | Hyperlink → Delete Hyperlink**.

alternatively
1. Point at the link in a Source Item while in Edit mode with the mouse pointer which will then become an arrow.
2. Right-click and select **Links → Hyperlink → Delete Hyperlink**.

The **Delete Hyperlink** dialog box appears:

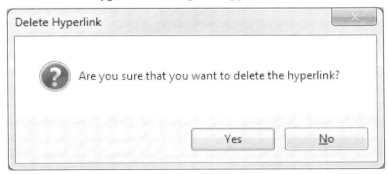

3. Confirm with **[Yes]**.

10. INTRODUCING NODES

By definition, a Node is a connecting point. In NVivo 10, Nodes are the primary tool for organizing and classifying source data. You can think of a Node as a 'container' of source material. Nodes can represent abstract concepts, such as topics, themes, and ideas. Nodes can also represent tangible concepts, such as people, places, and things. Remember, Nodes can represent anything you would find useful to organize and classify elements of your project. Some researchers know very early what kind of Nodes they will need to organize and categorize your data. You can create Nodes before you start to work with your source material. Other researchers may need to brainstorm organizational categories, concepts and structures 'on the fly' as they work through their source material. The way you work with Nodes varies largely depending on the methods used, the research situation and your personality.

Early on in any project, a good idea is to identify a few Nodes that you think will be useful. These early Nodes can be coded at as you work through your data for the first time. These early Nodes can be moved, merged, renamed, redefined or even deleted later on as your project develops.

NVivo also has developed a system for organizing and classifying both Source Items and Nodes, see Chapter 11, Classifications.

The terms *Parent Node*, *Child Node* and *Aggregate* are used when NVivo's Node system is described. A *Parent Node* is the next higher hierarchical Node in relation to its *Child Nodes*.

An *Aggregate*[1] means that a certain Node in any hierarchical level accumulates the logical sum of all its nearest Child Nodes. Each Node can at any point of time activate or deactivate the function Aggregate and with immediate effect. The Aggregate control is in the **New Node** dialog box or **Node Properties** dialog box.

Case Nodes and Theme Nodes

In our work, we find it useful to make a distinction between Theme Nodes and Case Nodes. Theme Nodes are containers based on themes, your ideas and insights about your project. Case Nodes are containers based on cases, the tangible elements of your project, like your participants or research settings. Importantly, Nodes have the ability

---

[1] *Aggregate* has an imperfection in that the number of references is calculated as the arithmetic sum of the Child Nodes' references, which instead should be the logic sum as some references are overlapping.

to be labeled with customized meta-data called Node Classifications. A Case Node is understood as a member of a group of Nodes which are classified with Attributes and Values reflecting demographic or descriptive data. Case Nodes can be people (Interviewees), places or any group of items with similar properties.

A Theme Node therefore represents a theme or a topic common to the whole project. Theme Nodes are often represented by a Node hierarchy. The research design of many qualitative studies is often based on the intersection between Case Nodes and Theme Nodes. This is obvious for the design of Node Matrices (see page 196) and Framework Matrices (see page 227).

Interviews are often very important in a qualitative study. Therefore it is important ta have a basic understanding of how an interview preferably is represented in an NVivo project. It is easy to let the interview become a Source Item and the interview person becomes a Case Node. The document is the Source and the person is the Node. Demographic characteristics (e.g., gender, age, education etc.) are then applied to the Case Node in the form of Attributes and Values. See Chapter 11, Classifications.

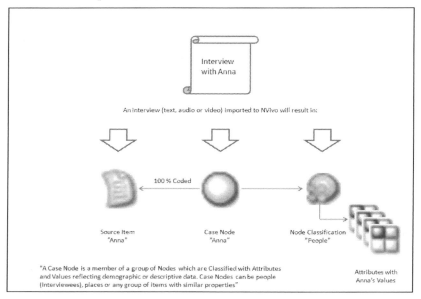

Creating a Node

Manually creating a new Node can be done in a number of ways.

1. Go to **Create | Nodes | Node**.
   Default folder is **Nodes**.
   Go to 5.

alternatively

1. Click [**Nodes**] in Area 1.
2. Select the **Nodes** folder in Area 2 or its subfolder.
3. Go to **Create | Nodes | Node**.
   Go to 5.

alternatively

3. Click on any empty space in Area 3.
4. Right-click and select **New Node...**
   or [**Ctrl**] + [**Shift**] + [**N**]

The **New Node** dialog box appears:

Tip: Some advice we offer coding newcomers is to record your thought processes in as much detail as possible when you are coding. The Description field of the new Node dialog box is an excellent place to capture why you have created that Node and how you think it relates to your coding hierarchy.

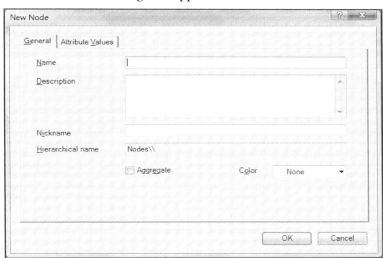

5. Type name (compulsory) and a description and a nickname (both optional), then [**OK**].

Here is a typical list view in Area 3 of some Nodes:

> **Tip:** Nicknames are only for nodes. A practical use when nodenames are very long. Use simple abbreviations. Must be unique within the project. Useful for the **Find** function and when coding with the Quick Coding Bar.

Building Hierarchical Nodes

As mentioned, Nodes can be organized hierarchically. As a result there are Node headings and subheadings in several levels of a coding hierarchy. Nodes can therefore form a sort of structured vocabulary, such as the MeSH (Medical Subject Headings) used by the Medline/PubMed article database.

Creating a Child Node

Assembling a Node hierarchy of Parent Nodes and Child Nodes is simple in NVivo:

1. Click [**Nodes**] in Area 1.
2. Select the **Nodes** folder in Area 2 or its subfolder.
3. Select the Node to which you want to create a Child Node.
4. Go to **Create | Nodes | Node**
   or right-click and select **New Node...**
   or [**Ctrl**] + [**Shift**] + [**N**].

The **New Node** dialog box now is shown.

5. Type a name (compulsory) and a description and a nickname (both optional), then [**OK**].

It is also possible to move Nodes within the list view, Area 3, with drag-and-drop or cut ([**Ctrl**] + [**X**]) and paste ([**Ctrl**] + [**V**]). When you drag one Node icon on top of another Node icon you create a child Node. You can also create a Child Node when you cut a Node, left-click a different Node, and then paste.

Here is a typical list view in Area 3 of some hierarchical Nodes:

Theme Nodes		
Name	Sources	References
Defining Volunteer Work	3	3
Interview Questions	9	135
Other Themes	0	0
Name	Sources	References
Motivation and Satisfaction	12	84
Name	Sources	References
Female Motivation	8	49
Male Motivation	7	35

Underlying items in the list can be opened or closed by clicking the + or – symbols, but also by using **View | List View | List View → Expand All (Selected) Nodes/ Collapse All (Selected) Nodes**. A useful function is showing Child Node Headers. When these headers are displayed you can modify the column widths. Apply **View | List View | List View → Child Node Headers** (toggling).

These menu options are also available by right-clicking in Area 3: **Sort By**, **List View** and **Expand/Collapse**.

Merging Nodes

Any Node can be merged into an existing Node. Merging two Nodes simply combines the content of one Node into another.

1. Cut or copy one or more Node(s).
2. Select a target Node.
3. Go to **Home | Clipboard | Merge → Merge Into Selected Node** or right-click and select **Merge Into Selected Node** or **[Ctrl] + [M]**.

In each case the **Merge Into Node** dialog box is shown:

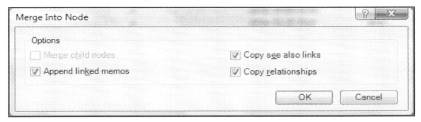

4. Select the applicable options, then click **[OK]**.

Alternatively, you can merge two (or more) Nodes into a new Node:

1. Cut or copy two or more Nodes.
2. Select the folder under which you want to place the new Node.
3. Go to **Home | Clipboard | Merge → Merge Into New Node...** or right-click and select **Merge Into New Node...**

alternatively

3. Select the parent node under which you want to place the new Node.
4. Go to **Home | Clipboard | Merge → Merge Into New Child Node** or right-click and select **Merge Into New Child Node...**

Tip: If you select *Append linked memos* a new memo will be created with the same name as the new node. If all merged nodes have memos the new memo will append the contents from all its memos.

The **Merge Into Node** dialog box appears:

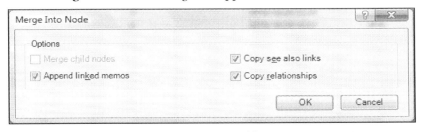

5. Select the applicable option(s), then click **[OK]**.

133

The **New Node** dialog box appears:

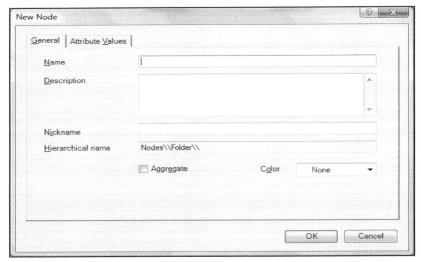

6   Type name (compulsory) and description and Nickname (optional), then **[OK]**.

Relationships

Relationships are Nodes that indicate that two Project Items (Source Items or Nodes) are related, such as the hypothesis *Poverty* **influences** *Public Health*. Data supporting that hypothesis could be coded at such relationship Node, which represents the relationship between the Nodes *Poverty* and *Public Health*.

Different relationship types are defined by the user and are stored under **[Classifications]** in the **Relationship Types** folder. Relationship Nodes are then created as associative, one way, or symmetrical (see below). The Relationships folder is not allowed to have subfolders and these Nodes cannot be arranged hierarchically. Classifications cannot be assigned to Relationships.

Creating a Relationship Type

Before creating relationships amongst your data, you must create some relationship types. In the above example (*Poverty* **influences** *Public Health*), the relationship type is titled **influences**.

1   Go to **Create | Classifications | Relationship Type**.
    Default folder is **Relationship Types**.
    Go to 5.

alternatively

1   Click **[Classifications]** in Area 1.
2   Select the **Relationship Types** folder in Area 2.

3   Go to **Create | Classifications | Relationship Type**.
    Go to 5.
alternatively
3   Click on any empty space in Area 3.
4   Right-click and select **New Relationship Type...**
    or **[Ctrl] + [Shift] + [N]**.

The **New Relationship Type** dialog box appears:

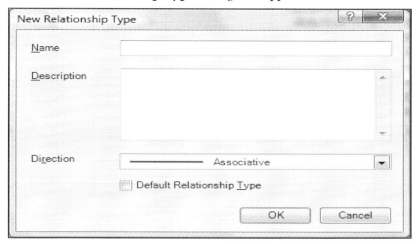

5   Select *Associative, One Way* or *Symmetrical* from the drop-down list at **Direction**.
6   Type a name (compulsory) and a description (optional), then **[OK]**.

The list view with Relationship Types in Area 3, may look like this:

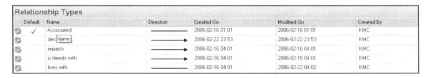

Creating a Relationship

Now that you have defined a relationship type, it's time to begin creating relationships between items in your NVivo project.

1   Go to **Create | Nodes | Relationships**.
    Default folder is **Relationships**.
    Go to 5.
alternatively
1   Click **[Nodes]** in Area 1.
2   Select the **Relationships** folder in Area 2.

3   Go to **Create | Nodes | Relationships**.
    Go to 5.
alternatively
3   Click on any empty space in Area 3.
4   Right-click and select **New Relationship...**
    or **[Ctrl] + [Shift] + [N]**.

The **New Relationship** dialog box appears:

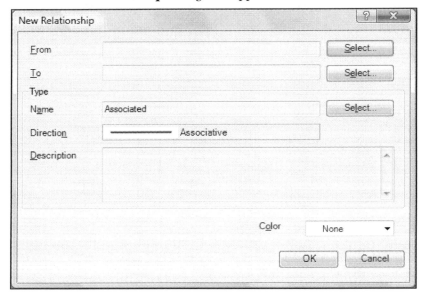

A Relationship defines a relation between two Project Items: Source Items or Nodes.

5   Use the **[Select...]** buttons to find the items that will be connected by this Node.

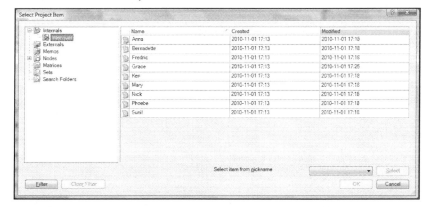

6   Select one From-item and one To-item. Confirm with **[OK]**.
7   Select a Relationship Type with **[Select...]** under **Type**.

The **New Relationship** dialog box will then look like this to represent a much more concrete relationship: that *Anna* **is married to** *Nick*:

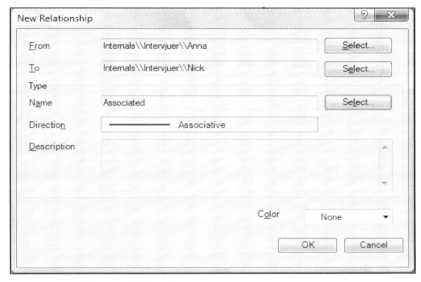

8   Confirm with [**OK**].

The list view with Relationships in Area 3 may look like this:

View a Relationship from a Related Item

1   Open an item in Area 3 that has a relationship.
2   Go to **View | Links → Relationships**
    which is a toggling function.

A new window will open and the relationship will show:

11. CLASSIFICATIONS

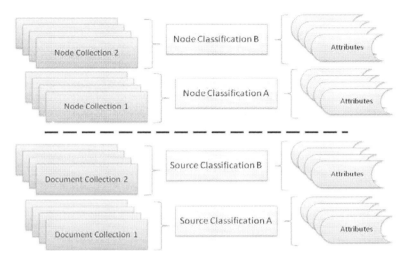

Node and Source Classifications

Nodes, Classifications and Attributes are related in the following way.

Source Items hold primary or secondary data. They can be text sources, media sources or picture sources.

Nodes represent a topic, a phenomenon, an idea, a value, an opinion, a case, or any other abstraction or tangible object thought to be important for the current study.

Attributes represent characteristics or properties of a Source Item or a Node which has or will have an impact when analyzing data. Each such attribute has a set of ***Values***. For example, if gender is your attribute, the only two possible values are mal or female.

Classifications are defined by NVivo as a collective name for a certain set of Attributes that will be assigned to certain Source Items or Nodes.

Classifications therefore fall into two types: Node Classifications and Source Classifications. We will explore how to create Classifications, how they are associated with Source Items and Nodes and how individual values are handled. Attributes cannot be created without the existence of Classifications. This chapter presents examples of how to create a Node Classification, but the procedures are similar for Source Classifications.

Node Classifications

An example: You are part of a study looking at the experiences of pupils, teachers, politicians and schools. There are reasons to create individual Nodes for each of these four groups:
- Attributes for pupils could then be: Age, gender, grade, number of siblings, social class.
- Attributes for teachers could then be: Age, gender, education, number of years as teacher, school subject.
- Attributes for politicians could then be: Age, gender, political preference, number of years as politician, other profile.
- Attributes for schools could then be: Size, age, size of the community, political majority.

Each of these four groups needs its own set of attributes, with each attribute requiring its own set of values. Each such set of attributes will collectively form a Node Classification.

Source Classifications

In NVivo, Classifications are also applied to Source Items with attributes and values. Source Classifications, for example, could be applied to certain interviews that may need attributes like the time of the inteview for longitudinal studies, place and other conditions. Source Classifications can also be applied to research that is the result of a literature review, with attributes like journal name, type of study, keywords, publication date, name of authors etc.

Creating a Classification

Although NVivo includes some default Classifications (e.g., the Node Classification *Person* and the Source Classification *Reference*), it is possible to create your own custom Classifications:

1. Go to **Create | Classifications | Node Classification**.
   Default folder is **Node Classifications**.
   Go to 5.

alternatively

1. Click **[Classfications]** in Area 1.
2. Select the **Node Classifications** folder in Area 2.
3. Go to **Create | Classifications | Node Classification**.
   Go to 5.

alternatively

3. Click on any empty space in Area 3.
4. Right-click and select **New Classification...**
   or **[Ctrl] + [Shift] + [N]**.

The **New Classification** dialog box appears:

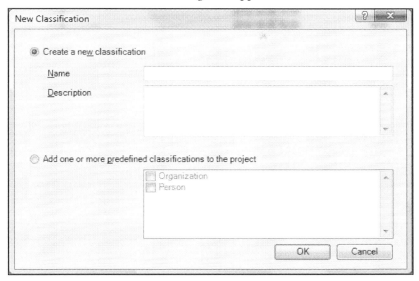

You can choose between creating your own new Classification or using one of NVivo's templates.

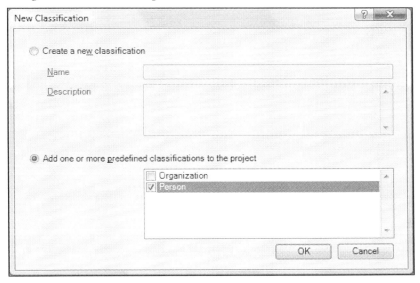

This example uses the template *Person*.
5   Click [**OK**].

The result is shown like this in Area 3:

The attributes that have been created from this template have initially no other values than *Unassigned* and *Not Applicable*.

The classification can easily be edited. You can create new attributes and you can delete those not needed.

Customizing a Classification

A Classification can easily be edited. You can also easily create new attributes and delete those not needed.

1. Click [**Classfications**] in Area 1.
2. Select the **Node Classifications** folder in Area 2.
3. Select a Classification in Area 3.
4. Go to **Create | Classifications | Attributes**
   or right-click and select **New Attribute...**
   or [**Ctrl**] + [**Shift**] + [**N**].

The **New Attribute** dialog box appears:

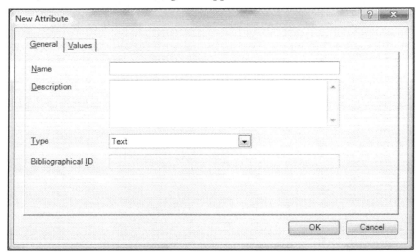

5. Type a name (compulsory) and a description (optional) and select the attribute type (Text, Integer, Decimal, Date/Time, Date, Time or Boolean), then [**OK**].

The data type field indicates what kind of data will constitute an Attribute's Values. There are seven data types: **Text** data includes any text content (e.g., profession); **Integer** data includes a number without a decimal place; **Decimal** data includes a number with a decimal place; **Time** data is the time in hours, minutes and seconds; **Date/Time** data is a combination of the calendar date and time; and **Boolean** data are binary pairs (e.g., yes or no, 0 or 1).

You can also decide which values belong to the attribute. Use either the **New Attribute** dialog box or the **Attribute Properties** dialog box, under the **Values** tab:

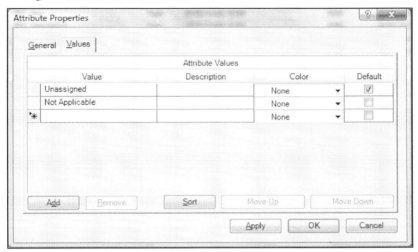

6   The **[Add]** button creates a text box in which you can type new values. Confirm with **[OK]**.

Finally, you need to assign the Classification to a Node.
1   Select one or several Nodes that shall be assigned a Classification.
2   Right-click and select **Classification** → <**Name of Classification**>.

Alternatively, if you only select *one* Node:
1   Select the Node that shall be assigned a Classification.
2   Right-click and select **Node Properties** or **[Ctrl]** + **[P]**.

The **Node Properties** dialog box appears:

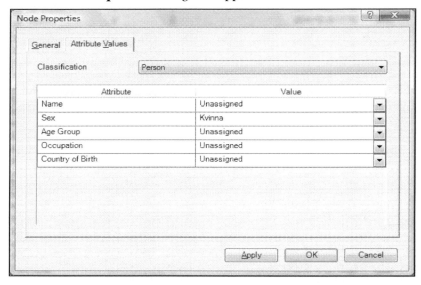

3   Use the **Attribute Values** tab. The **Classification** drop-down list will give you access to all Node Classifications in the project. Select your desired Node.
4   In the column *Value* you can use the drop-down list to set individual values to the current Node.
5   Confirm with [**OK**].

Working with the Classification Sheet

The overview of the attributes and values of Source Items or Nodes is called a Classification Sheet. This sheet is a matrix where rows are Source Items or Nodes and columns are Attributes. The cells contain the values.

While creating a Classification allows you to establish the Attributes and Values associated with classified items, the application of that metadata is done through the Classification Sheet. When a Classification Sheet is opened you can update values, along with sorting and filtering data.

1   Go to **Explore | Classification Sheets → Node Classification Sheets → <Name of Classification>**.

alternatively

1   Select a Classification in Area 3.
2   Right-click and select **Open Classification Sheet**.

alternatively

1   Select a classified Source Item or a classified Node in Area 3
2   Right-click and select **Open Classification Sheet**.

Below is a sample Classification Sheet. As you can see, each row is an item that has been classified with the Classification *Person*, each column is an Attribute, and each cell contains the attribute's attendant value:

	A : Age Group	B : Country	C : Ever done ...	D : Gender	E : Current par..	F : Education
1 : Anna	20-29	Aust	Yes	Female	Student	Tertiary
2 : Bernadette	60+	Aust	Yes	Female	Retired	Secondary
3 : Fredric	30-39	Aust	Yes	Male	Management Con	Tertiary
4 : Grace	20-29	Aust	Yes	Female	Marketing	Tertiary
5 : Kalle	60+	Aust	No	Male	Retired	
6 : Ken	50-59	Aust	Yes	Male	Retired	Secondary
7 : Mary	60+	Aust	Yes	Female	Retired	Secondary
8 : Nick	30-39	Aust	Yes	Male	IT	Tertiary
9 : Peter	30-39	Aust	No	Male	Marketing	Tertiary
10 : Phoebe	30-39	Aust	Yes	Female	Teacher	Tertiary
11 : Sunil	20-29	Aust	Yes	Male	Software Consult	Tertiary

Once you have your Classification Sheet open, there are a number of options for structuring, viewing and occluding aspects of your data.

Hiding/Unhiding Row Numbers (Toggling Function)
1. Open a **Classification Sheet**.
2. Go to **Layout | Show/Hide | Row IDs**
   or right-click and select **Row → Row IDs**.

Hiding Rows
1. Open a **Classification Sheet**.
2. Select one row or several rows that you want to hide.
3. Go to **Layout | Show/Hide | Hide Row**
   or right-click and select **Row → Hide Row**.

Hiding/Unhiding Rows with Filters
1. Open a **Classification Sheet**.
2. Click the 'funnel' in any column head
   or select a column and go to **Layout | Sort & Filter | Filter → Filter Row**.

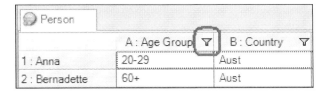

The **Classification Filter Options** dialog box appears:

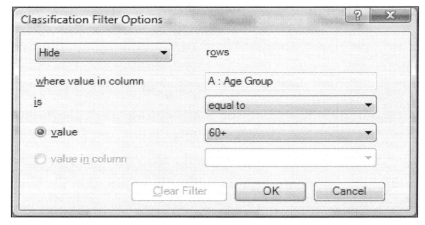

3   Select value and operator for hiding or unhiding. Confirm with **[OK]**. When a filter has been applied the funnel symbol turns *red*.

To reset a filter select **[Clear Filter]** in the **Classification Filter Options** dialog box.

Unhiding Rows
1   Open a **Classification Sheet**.
2   Select one row on each side of the hidden row that you want to unhide.
3   Go to **Layout | Show/Hide | Unhide Row**
    or right-click and select **Row → Unhide Row**.

Unhiding All Rows
1   Open a **Classification Sheet**.
2   Go to **Layout | Sort & Filter | Filter → Clear All Row Filters**
    or right-click and select **Row → Clear All Row Filters**.

Hiding /Unhiding Column Letter /Toggling Function)
1   Open a **Classification Sheet**.
2   Go to **Layout | Show/Hide | Column IDs**
    or right-click and select **Column → Column IDs**.

Hiding Columns
1   Open a **Classification Sheet**.
2   Select one column or several columns that you want o hide.
3   Go to **Layout | Show/Hide | Hide Column**
    or right-click and select **Column → Hide Column**.

Unhiding Columns
1. Open a **Classification Sheet**.
2. Select a column on each side of the hidden column that you want to unhide.
3. Go to **Layout | Show/Hide | Unhide Column**
   or right-click and select **Column → Unhide Column**.

Unhiding All Columns
1. Open a **Classification Sheet**.
2. Go to **Layout | Sort & Filter | Filter → Clear All Column Filters**
   or right-click and select **Column → Clear All Column Filters**.

Transposing the Classification Sheet (Toggling Function)
Transposing means that rows and columns switch places.
1. Open a **Classification Sheet**.
2. Go to **Layout | Transpose**
   or right-click and select **Transpose**.

Moving a Column Left or Right
1. Open a **Classification Sheet**.
2. Select the column or columns that you want to move. If you want to move more than one column they need to be adjacent.
3. Go to **Layout | Rows & Columns | Column → Move Left/Move Right**.

Resetting the Classification Sheet
1. Open a **Classification Sheet**.
2. Go to **Layout | Tools | Reset Settings**
   or right-click and select **Reset Settings**.

Exporting Classification Sheets

A Classification Sheet can be exported as a tab delimited text-file or an Excel spreadsheet:
1. Select the Classification Sheet in Area 3 that you want to export.
2. Go to **External Data | Export → Export Classfication Sheets...**

The **Export Classification Sheets** dialog box appears:

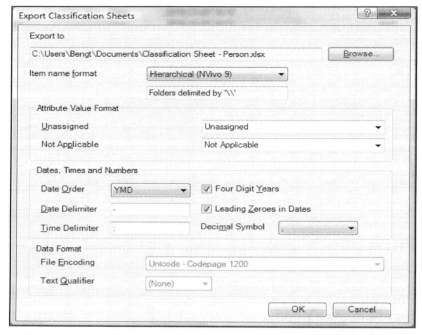

With **[Browse...]** you can decide name, file path and file format.

3   Confirm with **[OK]**.

Importing Classification Sheets

You can also import a Classification Sheet as a tab-delimited text-file or an Excel spreadsheet. All Nodes, Attributes and Values are created from the imported file if they do not exist already.

1   Go to **External Data | Import | Import Classification Sheets** or click **[Classifications]** in Area 1, click on any empty space in Area 3, right-click and select **Import Classification Sheets...**

Tip: An easy way to convert an Excel worksheet to text is:
1. Select the whole worksheet
2. Copy
3. Open Notepad
4. Paste into Notepad
5. Save with a new name

The **Import Classification Sheets Wizard – Step 1** appears:

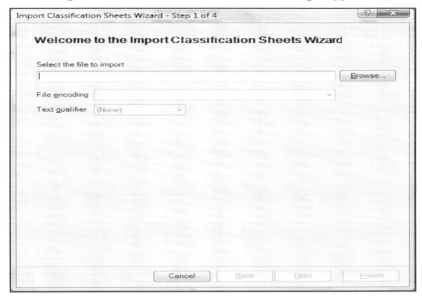

2   With **[Browse...]** you can find the file that you want to import.
3   Click **[Next]**.

The **Import Classification Sheets Wizard – Step 2** appears:

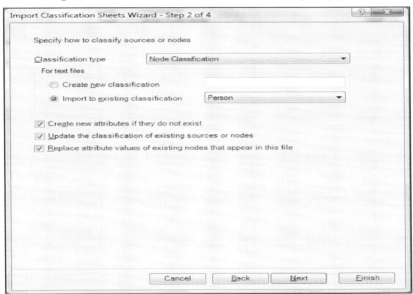

Here you decide if you want to create a new classification or use an existing one.

Create new attributes if they do not exist creates new attributes for the chosen classification.

Update the classification of existing sources or Nodes replaces the classification of the Source Items or Nodes that already exist in the location to be chosen.

Replace attribute values of existing Nodes that appear in this file determines if imported values shall replace the existing ones.

4   Click **Next**.

The **Import Classification Sheets Wizard** - **Step 3** appears:

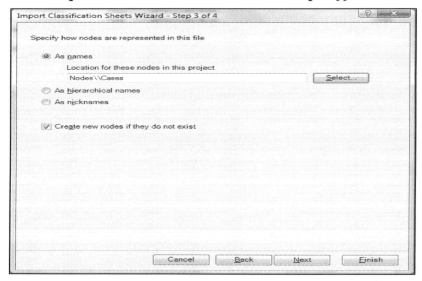

The option *As names* is selected when the first column of your file contains node names only. Requires that you use the **[Select]**-button to decide the location of the imported Nodes.

The option *As hierarchical names* is selected when the first column of your file contains the full hierarchical name, see page 21.

The option *As nicknames* is selected when the first column of your file contains the node nicknames, see page 132.

5   Use *As names* and the **[Select]**-button to decide the location of the Nodes when imported.

6   Click **[Next]**.

The **Import Classification Sheets Wizard – Step 4** appears:

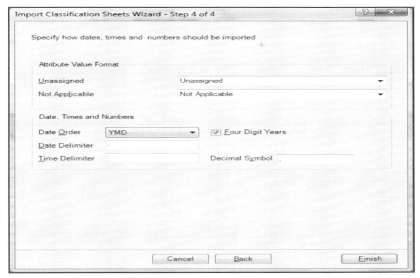

7   Decide the formats of unassigned values, dates, times and numbers.
8   Confirm with **[Finish]**.

The result, a Classification Sheet in NVivo, looks like this:

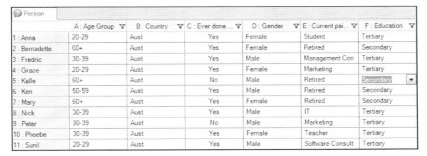

The Classification itself with its attributes can be displayed Area 3:

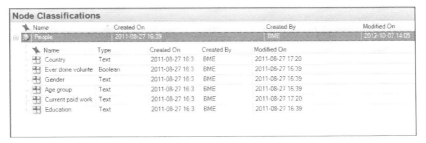

12. CODING

Coding is the act of assigning a portion of your source material to one of your Nodes. Coding can be carried out in two ways: *Manual Coding* (or just C*oding*) is conducted by the NVivo user; *Autocoding* is conducted by the NVivo software responding to pre-determined elements of source material.

The item being coded can be any piece of data, even something as small as a single word from a document or single frame from a video. Nodes are the set of conceptual terms or case information that you will code at. One usually says that you are *coding* a certain source element at a certain Node.

As arguably the most important function of qualitative data analysis software, NVivo offers a variety of methods for coding data:
- The Quick Coding Bar
- Drag-and-drop
- Right-click/Menus/Keyboard Commands
- Autocoding
- Range coding
- In Vivo coding
- Coding by Query

Here follows some basic definitions used both in the NVivo commands and in our instructions:

Code Sources implies that the entire content of a Source Item is coded.

Code Selection implies that a selected section in a Source Item is coded.

Code at Existing Nodes implies that the **Select Project Items** dialog box will appear for selection of one or several Nodes.

Code at New Node implies that the **New Node** dialog box will appear and you create and code at a new Node directly.

Code at Current Node implies that you code at the Node or the Nodes that were used most recently.

Code In Vivo implies that you instantly create a Node in the **Nodes** folder with the same name as the selected text data (max 256 characters).

The Quick Coding Bar

The Quick Coding Bar can be moved around on the screen or be positioned in the lower part of Area 4. You can toggle hiding/unhiding and docked/floating by going to **View | Workspace | Quick Coding** and the options **Hide, Docked** and **Floating**. Ideally, you will find the **Quick Coding Bar** useful enough to keep open each time you use NVivo – we do!

The **Quick Coding Bar** is active as long as a selection has been made in a Source Item or in a Node.

The drop-down list at **In** has three options: *Nodes, Relationships* and *Nicknames*. The first time in a new work session you normally select *Nodes* and then click on the first [...] button that now displays the **Select Location** dialog box. From here you can select among Node folders and parent Nodes.

We have explained Nodes and Relationships in this book, but Nicknames deserve some attention here. Nicknames are an opportunity for you to create 'shortcuts' to your most popular Nodes. For example, giving your 5 most popular Nodes nicknames allows you to efficiently access them from the **Quick Coding Bar** without needing to browse through your coding hierarchy. Nicknames can also be useful for creating shortened versions of Nodes with long names.

After selecting your Nodes, Relationships, or nickname, proceed to the drop-down list at **Code At**. This list contains all Nodes at the selected location in alphabetic order. You can also use the second [...] button that opens the **Select Project Items** dialog box thus giving access to all Nodes. You can select more than one Node to code at. You can also create a new Node by typing its name in the **Code At** text box at. The location of this new Node is determined by the setting in the left textbox **In**. Command **[Ctrl] + [Q]** positions the cursor in the right text box **Code At**, which will auto-complete Node names based on your typing – another shortcut.

The **Code At** drop-down also list saves the names of the last nine Nodes used during an ongoing work session. You find this list below a divider and in the order they were last used.

After you have selected your Nodes, the **Quick Coding Bar** can perform the following functions:

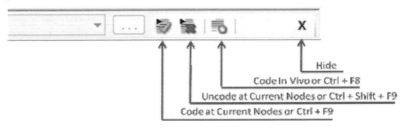

Drag-and-Drop Coding

Drag-and-drop coding is probably the fastest and easiest coding method. Using this method and a customized screen is, according to us, the best way to code your data.

1. Click **[Sources]** in Area 1.
2. Select the folder in Area 2 with the Source Item that you want to code.
3. Open the Source Item in Area 3 that you want to code.
4. Select the text or image that you want to code.
5. Click **[Nodes]** in Area 1 and select the folder with the Nodes that you want to code at.
6. With the left mouse button pressed, drag the selection from the Source Item to the Node that you want to code at.

Use Right Detail View to fully optimize your screen location. This view allows you to drag source data easily into a Node.

- **View | Workspace | Detail View → Right**
- **View | Workspace → Navigation View**

supports drag-and-drop (see page 47).

Menus, Right-Click, or Keyboard Commands

While we prefer drag-and-drop coding, you will no doubt find yourselves in situations where you need to code using another method.

Coding a Source Item

When you have a extant Node to connect your data with:

1. Click **[Sources]** in Area 1.
2. Select the folder in Area 2 with the Source Item that you want to code.
3. Select the Source Item or items in Area 3 that you want to code.
4. Go to **Analyze | Coding | Code Sources At →** <select>
   Existing Nodes   [Ctrl + [F5]]
   New Node   [Ctrl + [F6]]

alternatively

4. Right-click and select
   Code Sources → <select>
   Code Sources At Existing Nodes   [Ctrl + [F5]]
   Code Sources At New Node   [Ctrl + [F6]]
   Recent Nodes <select>

Coding a Source Item and Creating a Node
Not infrequently you will want to code data while creating a new Node at which to code that data:
1. Click **[Sources]** in Area 1.
2. Select the folder in Area 2 with the Source Item that you want to code.
3. Select the Source Item or items in Area 3 that you want to code.
4. Go to **Create | Items | Create As → Create As Node...**
   or right-click and select **Create As → Create As Node...**

The selected source(s) will be coded at a new Node and the Select Location dialog box will let you decide in what folder or under what parent Node the new Node will be created. Finally you need to give a name to the new Node.

Coding a Source Item and Creating Case Nodes
This function can be used when several Source Items need to be converted to Case Nodes. For example, you can create a list of Case Nodes if you have recently imported a number of interview transcripts. .
1. Click **[Sources]** in Area 1.
2. Select the folder in Area 2 with the Source Item that you want to code.
3. Select the Source Item or items in Area 3 that you want to code.
4. Go to **Create | Items | Create As → Create As Case Nodes...**
   or right-click and select **Create As → Create As Case Nodes...**

The selected sources will be coded at a new Node or Nodes and the **Select Location** dialog box will let you decide in what folder or under what parent Node the new Nodes will be located. One Node for each selected source will be created with the same name as the sources. The **Select Location** dialog box also makes it possible that you allocate one of the existing Node Classifications to the new Node or Nodes.

Coding a Selection from a Source Item

While Case Nodes will often pertain to entire source files, though not always, Theme Nodes often involve selections from a Source Item:

1. Click **[Sources]** in Area 1.
2. Select the the folder in Area 2 with the Source Item that you want to code.
3. Open the Source Item in Area 3 that you want to code.
4. Select the text or the section that you want to code.
5. Go to **Analyze | Coding | Code Selection At** → <select>
   Existing Nodes   [Ctr]l + [F2]
   New Node   [Ctrl] + [F3]

alternatively

5. Right-click and select
   Code Selection → <select>
   Code Selection At Existing Nodes   [Ctrl] + [F2]
   Code Selection At New Node   [Ctrl] + [F3]
   Code Selection At Current Nodes   [Ctrl] + [F9]
   Recent Nodes <select>

   Code In Vivo   [Ctrl] + [F8]

Uncoding a Source Item

As qualitative data coding is often an iterative process, sources may need to be uncoded.

1. Click **[Sources]** in Area 1.
2. Select the folder in Area 2 with the source that you want to uncode.
3. Select the Source Item or items in Area 3 that you want to uncode.
4. Go to **Analyze | Uncoding | Uncode Sources At** →
   Existing Nodes   [Ctrl] + [Shift] + [F5]

alternatively

4. Right-click and select
   Uncode Sources → <select>
   Uncode Sources At Existing Nodes   [Ctrl + [Shift] + [F5]
   Recent Nodes <select>

From the **Select Project Items** dialog box you select the node or nodes you want to uncode at.

Uncoding a Selection from a Source Item
1. Click **[Sources]** in Area 1.
2. Select the folder in Area 2 with the Source Item that you want to uncode.
3. Open the Source Item in Area 3 that you want to uncode.
4. Select the text or the section that you want to uncode.
5. Go to **Analyze | Uncoding | Uncode Selection At →**
   Existing Nodes   [Ctrl] + [Shift] + [F2]

alternatively

5. Right-click and select
   Uncode Selection → <select>
   Uncode Selection At Existing Nodes   [Ctrl] + [Shift] + [F2]
   Uncode Selection At This Node   [Ctrl] + [Shift] + [F3]
   Uncode Selection At Current Nodes   [Ctrl] + [Shift] + [F9]
   Recent Nodes <select>

Autocoding

At the onset of this chapter we said we would discuss two primary forms of coding, *Manual Coding* and *Autocoding*. This feature is based on the use of paragraph styles (Heading 1, Heading 2, etc.) to create a hierarchical Node structure. Autocoding codes the text under each heading under the name of the heading. If several documents are being auto coded at the same time or separately and they have the same structure of styles and headings then common Nodes are created automatically. A practical usage of this feature is when you apply a custom Word template with an established style set as a questionnaire for interviews. Autocoding can be applied to properly structured interview transcripts to, for example, code Source Item contents according to Case Nodes upon import:

> **Did you know?** Our website has custom autocoding Word templates to help structure your research data: www.formkunskap.com

1. Click **[Sources]** in Area 1.
2. Select the folder in Area 2 with the Source Items that you want to autocode.
3. Select the Source Item or items in Area 3 that you want to autocode.
4. Go to **Analyze | Code | Auto Code**
   or right-click and select **Auto Code...**

The **Auto Code** dialog box appears:

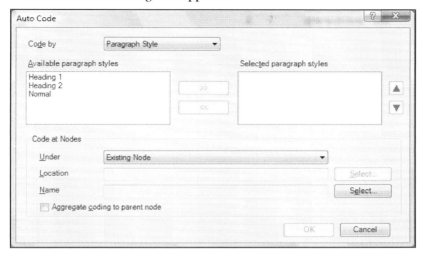

First, you need to decide which paragraph template should become the base for the new Node structure. NVivo will find all existing paragraph style templates in any Word document. The templates are selected with the [>>] button and are then transferred to the right textbox. The option *Existing Node* allows you to select the parent Node under which the new Nodes will be located. If you select *New Node* then you name the new Node and decide its location in a folder or under a parent Node. In either case underlying Nodes will be named after the text in the respective paragraph style (Heading 1, Heading 2 etcetera).

If you select **Code by** *Paragraph* each paragraph will be coded separately and the names of the Nodes will be the prevailing paragraph number.

5   Confirm with [**OK**].

- ♦ -

It is also possible to auto code transcripts of audio- or video-items. Suppose that we have an audio item and a transcript with two optional columns, Speaker and Organization:

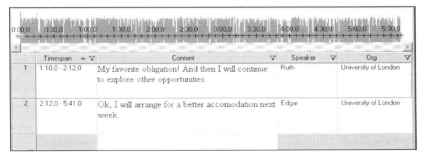

If you select **Code by** *Transcript Fields* autocoding such item will be based on these optional columns which will then create new Nodes named after the column contents:

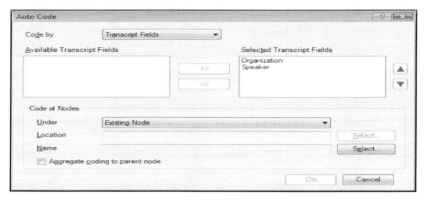

In this example *University of London* will become a Parent Node to the Child Nodes *Ruth* and *Edgar*.

Range Coding

Range coding is another principle for a rational coding of certain items. The basis for range coding is the paragraph number in a document, the row number in a transcript or picture log or the timespan in an audio- or video item.

The available options depend on the type of item that has been selected for range coding. The command is **Analyze | Coding | Range Code** and in this case only existing Nodes can be used to code at.

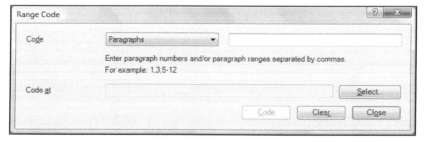

The coding takes place when you clicked at [**Code**].

In Vivo Coding

In Vivo coding is an established term used within qualitative research long before dedicated software existed. In Vivo coding creates a new Node from the selection of text and then, using the *In Vivo* command, the selected text (max 256 characters) will become the Node name. The new Node's location is always in the **Nodes** folder. Node name and location can be changed later.

1. Select the text you want to NVivo code.
2. Go to **Analyze | Coding | Code In Vivo**
   or right-click and select **Code In Vivo**
   or **[Ctrl]** + **[F8]**.

You can also use the Quick Coding bar described earlier in this chapter.

> **Tips**: Use In Vivo Coding like this: Select a Heading in your Source Item with a text that will become the Node name. Apply In Vivo Coding. Continue coding at this Node with the Quick Coding bar.

Coding by Queries

Queries can be instructed to save the result. The saved result is a Node and is instantly created when the query is run, see Chapter 13, Queries.

Visualizing your Coding

Opening a Node

1. Click [**Nodes**] in Area 1.
2. Select the **Nodes** folder in Area 2 or its subfolder.
3. Select the Node in Area 3 that you want to open.
4. Go to **Home | Item | Open → Open Node**
   or right-click and select **Open Node...**
   or double-click the Node in Area 3
   or **[Ctrl]** + **[Shift]** + **[O]**.

Each open Node is displayed in Area 4 and could therefore be docked or undocked. These windows always have a certain number of view mode tabs on its right side. If the Node has only been used to code text then the view mode tabs are: *Summary, Reference* and *Text*.

The *Reference* view mode is the default, automatically selected each time a Node is opened:

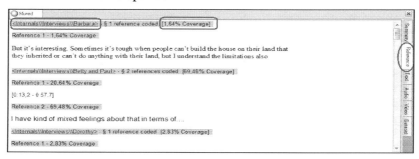

The link with the name of the Source Item opens in Area 4. You can also point at or select a section, go to **Home | Item | Open → Open Referenced Source** or right-click and select **Open Referenced Source**. When a Source Item is opened via a Node like this the coding at the current Node is highlighted.

References coded are coded segments (like a text-segment) of a source item.

Coverage means that the Node or a result of a query corresponds to a certain percentage of the whole Source Item that is coded measured in number of characters.

Hiding/Unhiding Reference to Source Items

1. Open a Node.
2. Go to **View | Detail View | Node → Coding Information**.
3. Uncheck *Sources, References* or *Coverage*.

The option *Sources* hides the reference to Source Items, coded sections and its coverage.

The option *References* hides information about each coded section and its coverage.

The option *Coverage* hides information about coverage of the Source Items and its coded sections.

The presentation can be modified in many ways by going to **View | Detail View | Node → Coding Context, Coding By Users, Coding Information, Coding Excerpt** and **Node Text**.

The *Summary* view mode displays all coded Source Items as a list of shortcuts. Each such shortcut can be opened with a double-click and the coded section is highlighted:

The *Text* view mode displays all coded text Source Items as thumbnails in the upper part of Area 4. Clicking on a thumbnail displays the coded section of that Source Item. Double-clicking the thumbnail opens the whole Source Item and the coded sections are highlighted:

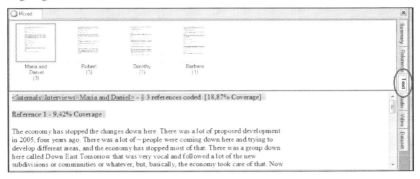

The *Audio* view mode provides a visual interface to easily listen to segments of coded audio:

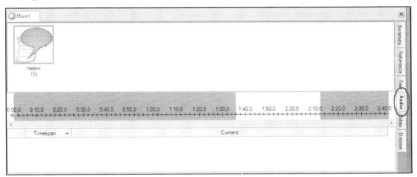

The *Video* view mode provides a visual interface to easily view segments of coded video:

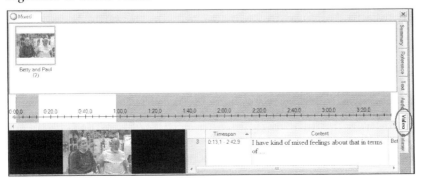

The *Picture* View mode provides an interface to view regions of coded image sources:

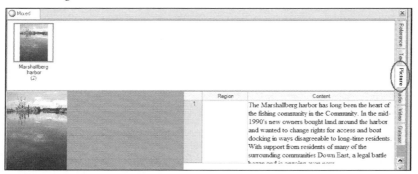

The *Dataset* view mode displays sections of coded dataset sources:

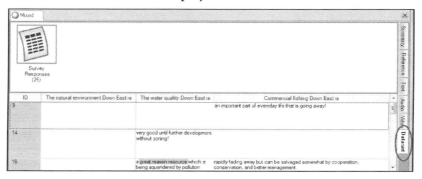

Viewing Coding Excerpt

1 Open a Node.
2 Go to **View | Detail View | Node → Coding Excerpt**.
3 Select *None, Start* or *All.*

The option *None*.

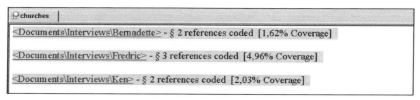

The option *Start*:

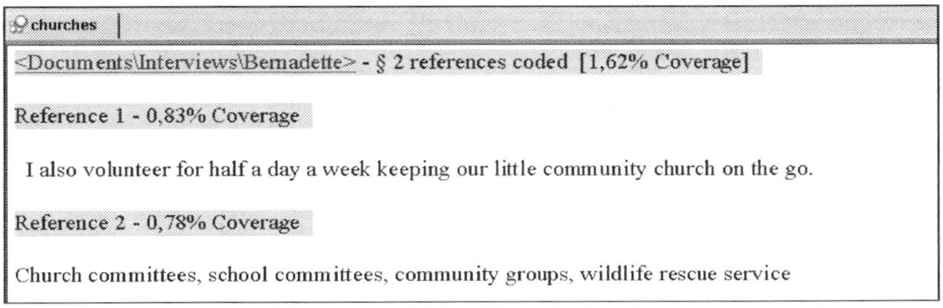

> churches
>
> <Documents\Interviews\Bernadette> - § 2 references coded [1,62% Coverage]
>
> Reference 1 - 0,83% Coverage
>
> I also volunteer for half a day a week keeping our little community church on the go.
>
> Reference 2 - 0,78% Coverage
>
> Church committees, school committees, community groups, wildlife rescue service

The option *All* is the default and has been shown above.

Viewing Coding Context
1. Open a Node.
2. Select the text or section that you want to show in its context.
3. Go to **View | Detail View | Node → Coding Context** or right-click and select **Coding Context**.
4. Select[2] *None, Narrow, Broad, Custom...* or *Entire Source*.

Example using the option *Broad* for a node coding a text source item:

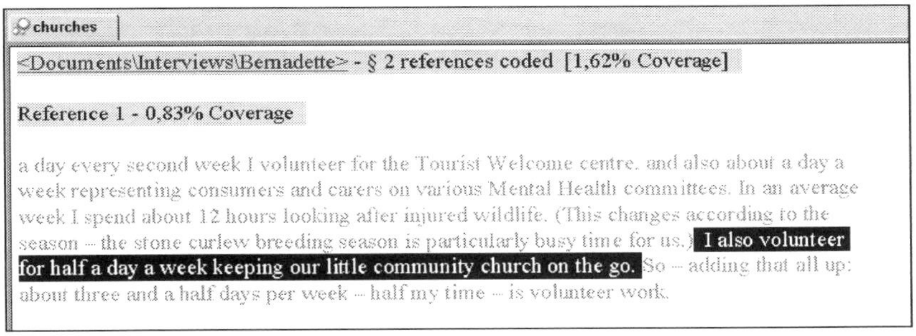

> churches
>
> <Documents\Interviews\Bernadette> - § 2 references coded [1,62% Coverage]
>
> Reference 1 - 0,83% Coverage
>
> a day every second week I volunteer for the Tourist Welcome centre, and also about a day a week representing consumers and carers on various Mental Health committees. In an average week I spend about 12 hours looking after injured wildlife. (This changes according to the season – the stone curlew breeding season is particularly busy time for us.) **I also volunteer for half a day a week keeping our little community church on the go.** So – adding that all up: about three and a half days per week – half my time – is volunteer work.

---

[2] The definition of *Narrow* and *Broad* is determined under **Application Options**, the **General** tab, see page 36. *Custom* can override these settings for any specific task.

Example using the option *Broad* for a node coding an Audio source. Playback is possible for the interval including the context.

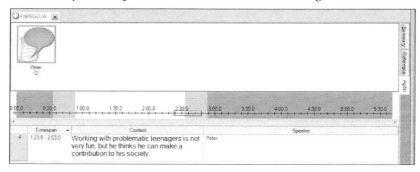

Example using the option *Narrow* for a node coding a Picture source.

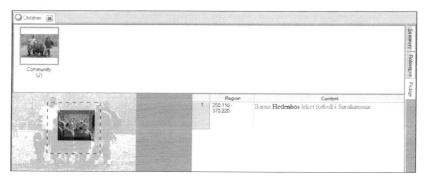

Highlighting Coding
The coded text or section in a Source Item can be highlighted in brownish color. Settings made are individual to Project Items and are temporarily saved during a work session, but are reset to none when a project is closed.

1. Go to **View | Coding | Highlight**.

There are several options:

None	Highlighting is off.
Coding For Selected Items...	Opens Select Project Items showing current Nodes, other Nodes are dimmed.
Coding for All Nodes	Highlights all Nodes that the Source Item is coded at.
Matches For Query	Highlights the words used by Text Search Queries.
Select Items...	Opens Select Project Items and you can modify selection of Nodes.

Coding Stripes

The open document, memo, or Node can be made to show the current coding as colored vertical stripes in a separate right hand window. Coding stripes are shown in Read-Only mode or in Edit mode. Using the Refresh link on top of the window recovers colors and functions of the stripes.

1. Go to **View | Coding | Coding Stripes** (There are several options):

None	Coding Stripes are off.
Selected Items...	Is active when coding stripes have been selected.
Nodes Most Coding	Shows the Nodes that are most coded at.
Nodes Least Coding	Shows the Nodes that are least coded at.
Nodes Recently Coding	Shows the Nodes that are recently coded at.
Coding Density Only	Shows only the Coding Density Bar and no Nodes.
Selected Items	Opens Select Project Items showing current Nodes, other Nodes are dimmed.
Show Items Last Selected	Shows the Nodes that were last opened.
Number of Stripes...	Selects the number of stripes (7 – 200).

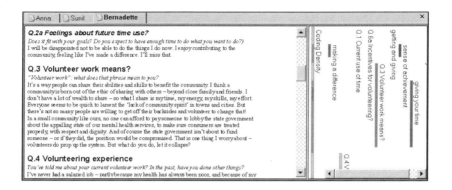

What can you do with Coding Stripes?

When you point and right-click at a certain coding stripe the following options will show: **Highlight Coding, Open Node..., Uncode, Hide Stripe, Show Sub-Stripes, Hide Sub-Stripes** and **Refresh**.

A click on the coding stripe highlights the coded area and double-click opens the Node.

By pointing at a coding stripe the Node name is shown. By pointing at the Coding Density Bar all Node names are shown that are coded at near the pointer.

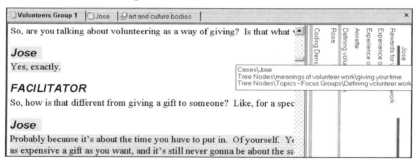

Color Marking of Coding Stripes

The colors of the coding stripes are automatically selected by NVivo. You can also use a custom color scheme, see page 23.
1. Show coding stripes using any of the above options.
2. Go to **View | Visualization | Color Scheme → Item Colors**.

Nodes without individual colors in the custom color scheme will be shown without any color.

Printing with Coding Stripes
See Chapter 5, section Printing with Coding Stripes (page 81).

Charts
Charts are graphics that easily and clearly can illustrate how sources have been coded. The generic way to create Charts is using the Chart Wizard.

    1   Go to **Explore | Visualizations | Chart**.

The **Chart Wizard – Step 1** appears:

    2   Click [**Next**].

The **Chart Wizard – Step 2** is shown and the options are:

Coding (Create a chart for coding) and the alternatives are:
    Coding for a source
    Coding by attribute value for a source
    Coding by attribute value for multiple sources
    Coding for a Node
    Coding by attribute value for a Node
    Coding by attribute value for multiple Nodes

Sources (Create a chart for sources) and the alternatives are:
  Sources by attribute value for an attribute
  Sources by attribute value for two attributes

Nodes (Create a chart for Nodes) and the alternatives are:
  Nodes by attribute value for an attribute
  Nodes by attribute value for two attributes

Option	Comments
Coding for a source	Compare the Nodes used to code a particular source. For example, chart any source to show the Nodes which code it by percentage of coverage or number of references.
Coding by Node attribute value for a source	Show coding by Node attribute value for a source. For example chart a source to show coding by one or more Node attribute values.
Coding by Node attribute value for multiple sources	Show coding by Node attribute value for multiple sources. For example chart two or more sources to show coding by one or more Node attribute values.
Coding for a Node	Look at the different sources that atre coded at a Node. For example, chart any Node to see which sources are coded at the Node and their corresponding percentage of coverage.
Coding by Node attribute value for a Node	Show coding by attribute value for a Node. For example, chart a Node to show coding by one or more attribute values.
Coding by Node attribute value for multiple Nodes	Show coding by attribute value for multiple Nodes. For example, chart two or more Nodes to show coding by one or more attribute values.
Sources by attribute value for an attribute	Display sources by attribute value for an attribute. For example chart an attribute to see how the sources which have that attribute are distributed across the attribute values.
Sources by attribute value for two attributes	Display sources by attribute value for two attributes. For example chart two attributes to see how the sources which have those attributes are distributed across the attribute values.
Nodes by attribute value for an attribute	Display Nodes by attribute value for an attribute. For example chart an attribute to see how the Nodes which have that attribute are distributed across the attribute values.
Nodes by attribute value for two attributes	Display Nodes by attribute value for two attributes. For example chart two attributes to see how the Nodes which have those attributes are distributed across the attribute values.

3 Click **[Next]**.

The **Chart Wizard** – **Step 3** appears:

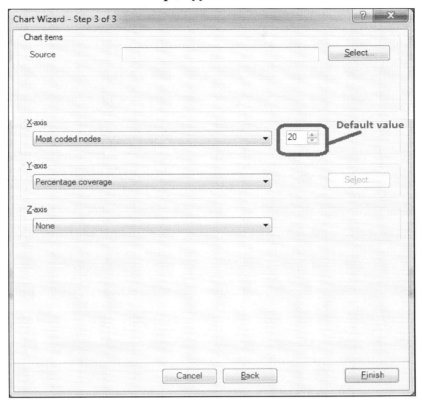

4   Use the **[Select]** button to choose the item that you will visualize, then **[Finish]**.

The result can be like this:

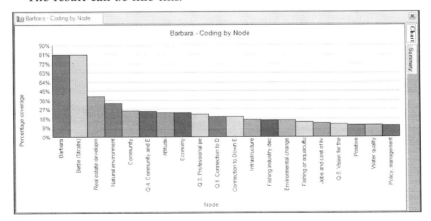

The ribbon menu **Chart** opens and makes it possible to modify the formatting, zooming and rotating. By going to **Chart | Type** the following drop-down menu shows:

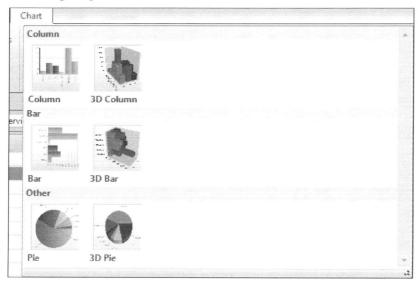

Here you can choose from creating a column, a bar or a pie diagram.

What else can you do with a Chart?
- Hover over the chart and you can read the exact coverage.
- **Chart <Source Item> Coding**:
  Double-click on a bar and you will open the selected Node in a Highlight mode on for the current Source Item.
- **Chart Node Coding**:
  Double-click on a bar and you will open the selected source in a Highlight mode on for the current Node.

The *Summary* tab displays a list with Nodes and their coverage:

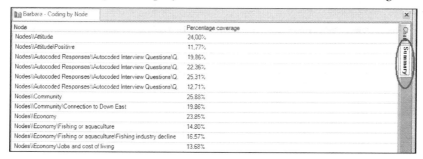

- ♦ -

During a work session you can also start from Area 3:
1. Select the item in Area 3 that you want to visualize.
2. Go to **Explore | Visualizations | Chart → Chart <Item type> Coding**
   or go to **Explore | Visualizations | Chart → Chart <Item type> by Attribute Value**.

alternatively
2. Right-click and select **Vizualize → Chart <Item type> Coding** or **Chart <Item type> by Attribute Value**.

The graphic is then shown directly when you select **Chart <Item type> Coding** or the **Chart Options** dialog box appears (same box as the **Chart Wizard – Step 3**). From there you proceed as above. These charts cannot be saved as items in the project but can be exported in the following file formats: .JPG, .BMP, or .GIF.

1. Create a Chart.
2. Go to **External Data | Export | Export → Export Chart**
   or right-click in the graph and select **Export Chart**
   or **[Ctrl] + [Shift] + [E]**.
3. Use the file-browser to decide filename and location, then click **[Save]**.

If you want to display more than 20 items in a chart then:
1. Open a chart.
2. Go to **Chart | Options | Select Data**.

The **Chart Options** dialog box appears:

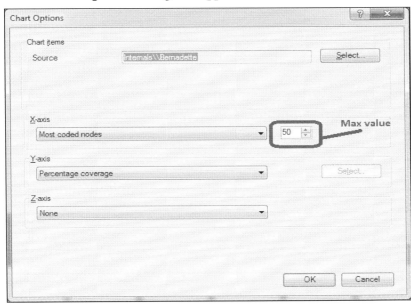

3. Change the settings, then **[OK]**.

Viewing a Node that Codes a PDF Source Item

The PDF Source
A coded PDF showing coding stripes can for example look like this (the content of bookmark panel depends on the original PDF and can be hidden or unhidden by going to **View | Window | Bookmarks**):

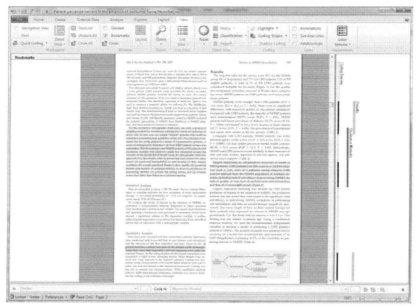

When you need to code a complete PDF document with any of the commands '**Code Sources at <Node>**' or '**Create As Node**' or '**Create As Case Nodes**' the number of references in that Node is calculated like this:

All text is one reference and each page is a region.

The Node
A Node that codes a PDF item will show the PDF in the Summary tab, in the Reference tab and in the specific PDF tab as follows:

The **Summary** tab will show the PDF as any other shortcut in the list of coded items.

The **Reference** tab will show coded text as plain text and coded region as coordinates:

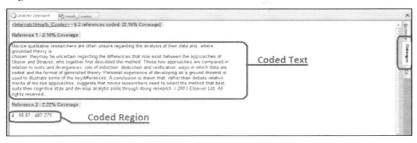

The **PDF** tab will show coded text or region as clear windows in the original PDF layout. Only coded pages with show:

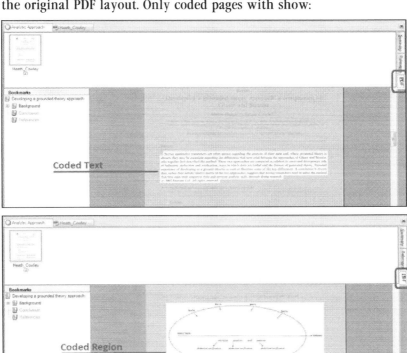

13. QUERIES

This chapter is about how to create and run various kinds of queries. In our experience, new NVivo users are sometimes intimidated by Queries – many types exist and using them effectively can take some practice. Remember, although not every query type will be right for project, every query type requires similar elements of foundational input. You will find queries increasingly simple once you read this chapter and learn the key features of a Query.

When you create a query, you first decide whether it will become a new NVivo item within the Queries folder (Area 1). The option *Add to Project* is available in the dialog boxes **New <Type> Query** and **<Type> Query Properties**. Add to Projcet lets you type a name of the query and it will be saved for future use. The saved queries respond the same as other NVivo items – they can be copied, pasted, and moved into folders. Query items open into query dialog boxes where you can adjust the settings of each query. Importantly, you will need to **Run** a query before you will see any search results; queries can be created without being **Run**.

You can construct simple queries that find certain items or text elements. You can also construct complex queries that combine search words and Nodes or that combine several Nodes. The results of queries based on search words and Nodes can generate new Nodes, sets or data visualizations like Word Clouds, or both. You can also merge query results with existing Nodes.

NVivo offers seven different query types, Text Search Queries, Coding Queries, Matrix Coding Queries, Word Frequency Queries, Compound Queries, Group Queries, and Coding Comparison Queries, which we discuss in Chapter 22, Collaborating with NVivo 10. Saving a query, editing a query, moving a query to another folder, deleting a query and previewing or saving results are dealt with in the next chapter, Common Query Features.

Word Frequency Queries

Word Frequency Queries makes it possible to make a list of the most frequent words in selected items: Source Items, Nodes etc.
1. Go to **Query | Create | New Query → Word Frequency...**
   Default folder is **Queries**.
   Go to 5.

alternatively
1. Click **[Queries]** in Area 1.
2. Select **Queries** folder in Area 2 or its subfolder.
3. Go to **Query | Create | New Query → Word Frequency...**
   Go to 5.

alternatively
3. Click at an empty space in Area 3.
4. Right-click and select **New Query → Word Frequency...**

The **Word Frequency Query** dialog box appears:

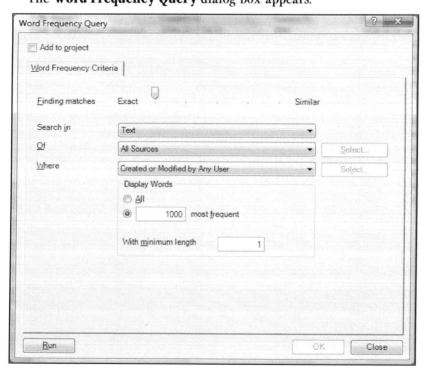

The **Finding matches** slider is the same as we will describe under **Text Search Queries** (see page 183).
5. When choosing *Selected Items* from the **Of** drop-down list and then **[Select...]** the **Select Project Items** dialog box appears and is used like a Text Search Query.

6   When items and other options have been decided, then click
    [**Run**].

The result may look like this, with the *Summary* tab open by default:

Word	Length	Count	Weighted Percentage (%)
volunteer	9	308	1.55
work	4	287	1.44
time	4	280	1.41
what	4	255	1.28
have	4	247	1.24
about	5	238	1.20
like	4	232	1.17
volunteering	12	185	0.93
your	4	177	0.89
volunteers	10	172	0.87
people	6	170	0.86
think	5	165	0.83
some	4	129	0.65
just	4	125	0.63

Select *one* word (it is not possible to select more than one), right-click and the following options appear:

- *Open Node Preview* (or double-click or key command [**Ctrl**] + [**Shift**] + [**O**])
  Opens like any Node with search words and synonyms highlighted with Narrow Coding Context (5 words).
- *Run Text Search Query*
  The **Text Search Query** dialog box is shown with the search words and synonyms transferred to the search criteria. The options Selected Items is inherited from the **Word Frequency Query** dialog box. The dialog box can be edited before you run it. See also page 181 on what you can do with Text Search Queries.
- *Export List...*
- *Print List...*
- *Create As Node...*
  Creates a Node with the search words and synonyms and a Narrow Coding Context (5 words). The Context Setting is retained in the Nodes folder or its subfolder during the ongoing work session.
- *Add to Stop Words List*[3]

---

[3] Alternatively: Go to **Query | Actions | Add to Stop Word List**

The *Tag Cloud* tab displays a custom tag cloud based on your query:

The Tag Cloud tab displays up to 100 words. The size of the words reflects their frequency. Words are sorted alphabetically and include stemmed words and synonyms if the Word Frequency Query is set accordingly. Click on a word and a Text Search Query is created and runs with results displayed as a Node preview.

The *Tree Map* tab displays a custom tree map based on your query:

The Tree Map tab displays up to 100 words. The size of the area of each element reflects the frequency of the word. Click on a word and a Text Search Query is created and run with the results displayed as a Node preview. For more on Tree Maps, see page 311.

The *Cluster Analysis* tab displays a custom cluster analysis based on your query:

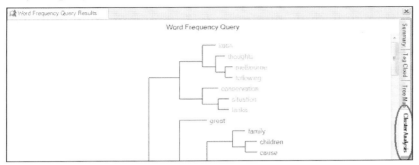

The Cluster Analysis tab displays up to 100 words. Words that co-occur are clustered together. When this tab has been selected the Ribbon menu **Cluster Analysis** is shown and you can choose between 2D Cluster Map, 3D Cluster Map, Horizontal Dendrogram or Vertical Dendrogram. With **Cluster Analysis | Options | Select Data** you can choose the metric coefficient. Checking **Cluster Analysis | Options → Word Frequency** when viewing 2D or 3D Cluster Maps lets you present the size of symbols reflecting its occurence. **Cluster Analysis | Options → Clusters** is any number between 1 and 20 (10 is default) representing the number of colors used in the cluster diagrams.

Tag Clouds, Tree Maps, and Cluster Analysis can also be used like this: Select a word in the graph, right-click and the menu is similar to the one explained on page 179. Only the Cluster Analysis har two unique alternatives: Copy (the whole graph) and Select Data (Pearson, Jaccard's or Sørensen's coefficients).

For more on Cluster Analysis, see page 307.

Text Search Queries

Text Search Queries search for certain words or phrases in a set of items:
    1   Go to **Query | Create | New Query → Text Search...**
        Default folder is **Queries**.
        Go to 5.
alternatively
    1   Click [**Queries**] in Area 1.
    2   Select **Queries** folder in Area 2 or its subfolder
    3   Go to **Query | Create | New Query → Text Search...**
        Go to 5.
alternatively
    3   Click on an empty space in Area 3.
    4   Right-click and select **New Query → Text Search...**

The **Text Search Query** dialog box appears:

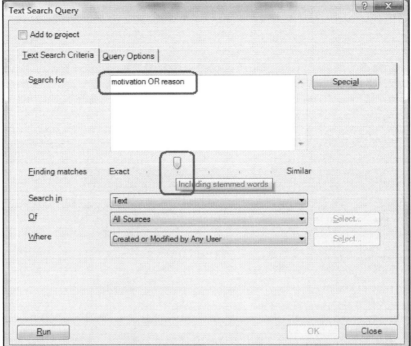

5   Type the search word or the search criteria in the **Search for** text box, for example `'motivation OR reason'`. Move the slider **Finding matches** over the option *Including stemmed words*, this way the query searches words with same stem as the typed search words. (English, French, German, Portuguese and Spanish only).

When several words are typed in a sequence, e.g. `ADAM EVE`, the search is made as an OR-combination and when the words are surrounded by quotes, `'ADAM EVA'`, an exact phrase search is run.

The slider **Finding matches** has five options:

Position	Result	Example
Exact match	Exact matches only	sport
Including stemmed words	Exact matches Words with the same stem	sport, sporting
Including synonyms	Exact matches Words with same stem Synonyms[4] (words with a very close meaning)	sport, sporting, play, fun
Including specializations	Exact matches Words with same stem Synonyms[1] (words with a very close meaning) Specializations (words with a more specialized meaning)	sport, sporting, play, fun, running, basketball
Including generalizations	Exact matches Words with same stem Synonyms[1] (words with a very close meaning) Specializations (words with a more specialized meaning—a 'type of') Generalizations (words with a more general meaning)	sport, sporting, play, fun, running, basketball, recreation, business

All settings work for NVivo's text content languages. The text content language options are available when you go to **File → Info → Project Properties**, the **General** tab: *Text Content Language* (see page 51). If this setting is made for *Other* then only 'Exact match' can be used but can be combined with the conventional operators under **[Special]** which offers the following optional search functions:

Option	Example	Comment
Wildcard ?	ADAM?	? represents *one* arbitrary character
Wildcard *	EVA*	* represents *any number* of arbitrary characters
AND	ADAM AND EVA	Both words must be found
OR	ADAM OR EVA	Either word must be found
NOT	ADAM NOT EVA	Adam is found where Eve is not found
Required	+ADAM EVA	Adam is required but Eve is also found
Prohibit	-EVA ADAM	Adam is found where Eve is not found
Fuzzy	ADAM~	Finds words of similar spelling
Near...	"ADAM EVA"~3	Adam and Eve are found within 3 words from each other
Relevance...	ADAM^2EVA	Adam is 2 times as relevant as Eve is

6   Confirm with **[Run]**.

---

[4] Each content language has its own built-in, non-editable synonym list.

The format of the result depends on the settings made under the **Query Options** tab in the dialog box **Text Search Query** (see page 211).

After you run a Text Search Query, the *Preview Only* option displays a list of shortcuts in Area 4 and can look like this. These shortcuts contain the search results within a given Source Item. The Summary tab is default:

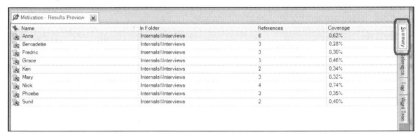

The list of shortcuts can easily be sorted by clicking on the column head. When you double-click on such shortcut the item will open and the search words are highlighted:

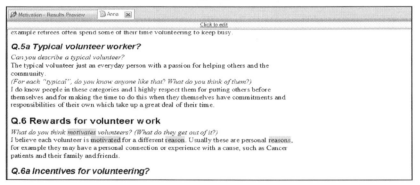

Creating a Set
You might find it useful to combine results from your search into a Set:
1   Select the shortcuts that you want to create as a set.
2   Go to **Create | Collections | Set → Create As Set...**
    or right-click and select **Create As → Create As Set...**
3   Type a name of the new set and confirm with **[OK]**.

alternatively, if you already have a set:
1   Select the shortcuts that you want to add to a set.
2   Go to **Create | Collections | Set → Add To Set...**
    or right-click and select **Add To Set...**
3   Select Set in the **Select Set** dialog box.
4   Confirm with **[OK]**.

Creating a Node
You can also combine results from your search into a new Node:
1. Select the shortcuts that you want to create as a Node.
2. Right-click and select **Create As → Create As Node...** (selected shortcuts will be merged into one new Node) or right-click and select **Create As → Create As Case Nodes...** (selected shortcuts will each become a new Node).
3. In the **Select Location** dialog box you must determine where the new Node or Nodes shall be located.
4. Type a name for the new Node. When Case Nodes are created they will inherit the names of the sources. The **Select Location** dialog box makes it optional that you allocate one of the existing Node Classifications to the new Case Node(s). Confirm with [**OK**].

Saving Search Results
1. Select the shortcuts that you want to create as a Node.
2. Right-click and select **Store Query Results** (all shortcuts will be merged into one new Node) or right-click and select **Store Selected Query Results** (selected shortcuts will be merged into one new Node)

The **Store Query Results** dialog box is shown.
3. Determine the name and location of the new Node.

The *Reference* view mode displays 5 words on each side of the search word (Coding Context) and otherwise the view options are the same as for an open Node (see Chapter 12, section Visualizing your Coding, page 161):

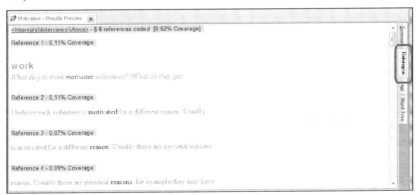

The *Text* view mode is also identical as for an open Node (see page 163):

Word Trees

The *Word Tree* view mode is a feature for Text Search Queries that visualizes how a word occurs within a corpus of sentences. This is one of our favorite NVivo visualizations. Remember, to generate a Word Tree you need to ensure query options set for *Preview* and that Spread Coding is off:

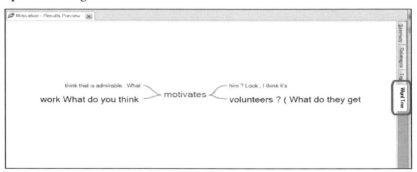

A new Ribbon menu, **Word Tree**, now appears and there you can find a list called Root Term. This list is sorted by frequency, and displays words resulting from the placement of the **Finding matches** slider. Each selected Root Term creates a new Word Tree. You can also decide the number of words (Context Words) that surrounds a Root Term.

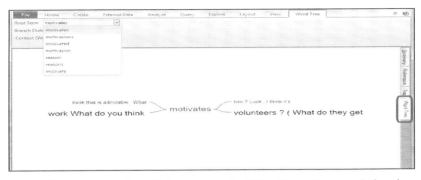

Finally, you can also click any word of the Word Tree and the the whole branch will be highlighted. Double-clicking a selected branch opens a Node preview. Clicking the root term will highlight all branches of the Word Tree. You can also select a branch, right-click and the following menu appears: Run Text Search Query (similar to double-clicking the branch), Export Word Tree, Print and Copy. A full Word Tree can also be exported as a low-resolution image. Unfortunately, at present NVivo does not contain functionality for exporting high-resolution images. But we have requested this function and we are hopeful for the future.

Coding Queries

Coding Queries are advantageous when you have advanced your project's structure in such a way that you can acquire project insights via complex queries.

1 Go to **Query | Create | New Query → Coding...**
   Default folder is **Queries**.
   Go to 5.

alternatively

1 Click **[Queries]** in Area 1.
2 Select the **Queries** folder in Area 2 or its subfolder.
3 Go to **Query | Create | New Query → Coding...**
   Go to 5.

alternatively

3 Click at an empty space in Area 3.
4 Right-click and select **New Query → Coding...**

The **Coding Query** dialog box appears:

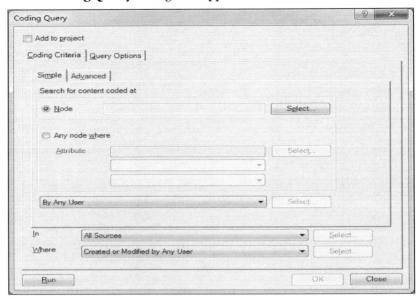

5   Go to the **Coding Criteria** tab and select either the **Simple** tab or the **Advanced** tab.

Coding Criteria Tab → Simple Tab

Example: What motivates people in the age group 30 - 39 to do volunteering? Let's first limit the search to the Node *Personal Goals*.

1   Select **Coding Criteria** → **Simple** tab.
2   Choose *Selected Items* from the drop-down list **In**.
3   Click the [**Select...**] button.

The **Select Project Items** dialog box appears:

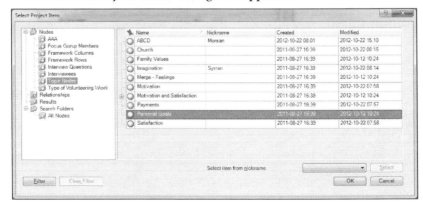

In the left part of the dialog box the whole folder structure of the project is shown and in the right part you can find the items (Source Items, Nodes etc.) that belong to the current folder.

4   We select the Node *Personal Goals*. When we have confirmed with [**OK**] we return to the **Coding Query** dialog box.

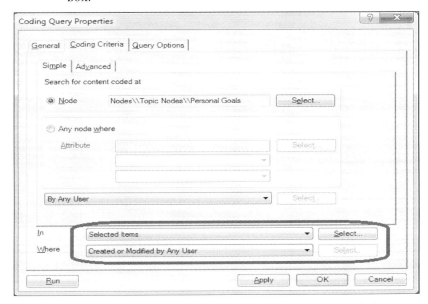

If we search in All Sources we could as well have opened the Node *Personal Goals* directly. The options **In** and **Where** in the lower part of the dialog box makes it possible to limit the search to Selected Items, Selected Folders or by any Users.

The format of the result depends on the settings of the **Query Options** tab in the dialog box **Coding Query** (see page 211).

Criteria Tab → Advanced Tab

The **Simple** tab and the **Advanced** tab are independent of each other. The **Advanced** tab makes more complex search criteria possible. In this example, we want to limit the search to *Females* in the *Age group 30 - 39* and then search the Nodes *Personal Goals* and *Family Values*.

1   Select **Coding Criteria → Advanced** tab.

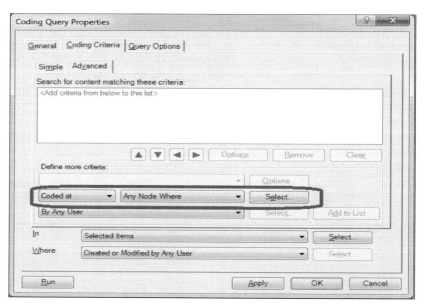

2 Under **Define more criteria** choose the option *Coded at* and *Any Node Where*, then the [**Select...**] button.

The **Coding Search Item** dialog box appears:

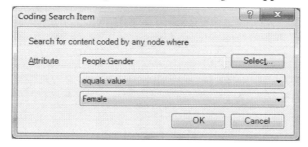

3 Select Attribute and Value *Gender / equals value / Female*.
4 Confirm with [**OK**] and then click [**Add to List**] in the **Coding Query** dialog box.

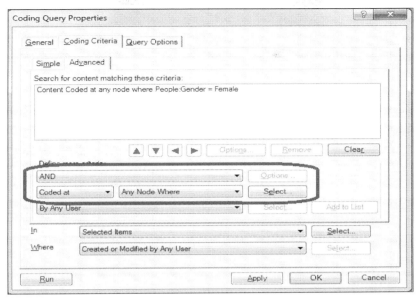

5   Return to the section **Define more criteria**, select the operator[5] *AND, Coded at* and *Any Node Where*.
6   Use the **[Select...]** button and select *Age group / equals value / 30 - 39* in the **Coding Search Item** dialog box.

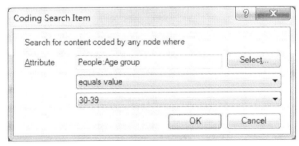

7   Confirm with **[OK]** and then click **[Add to List]** in the **Coding Query** dialog box.

---

[5] See page 213 onward for explanations of the other operators on this drop-down list.

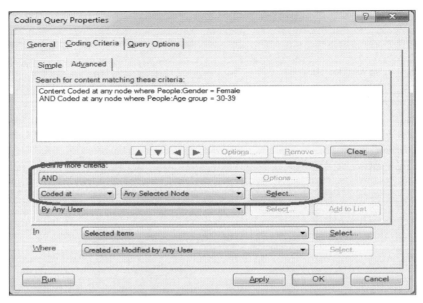

8   Return to the section Define more criteria, select the operator[6] *AND, Coded at* and *Any Selected Node Where*. With the **[Select]** button you select the two Nodes *Personal Goals* and *Family Values.*

9   Confirm with **[OK]** and then click **[Add to List]** in the **Coding Query** dialog box.

---

[6] See page 213 onward for explanations of the other operators on this drop-down list.

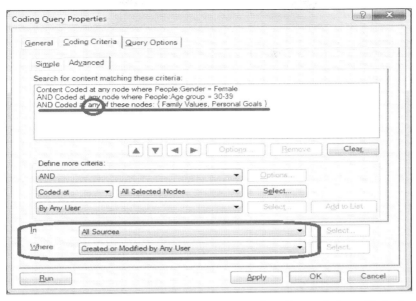

In this last criterion, it is important to select 'Any of these Nodes' which means a logical OR between the selected Nodes. We are now searching *All Sources* as the criteria limits anyway.

10   Click **[Run]** in the **Coding Query** dialog box.

The format of the result depends on the settings of the **Query Options** tab in the dialog box **Coding Query (Properties)**. See more on this on page 211.

When you have run a Coding Query with *Preview Only* the result looks like under Text Search Queries, page 184 and onwards, with two exceptions: The Word Tree tab is not included and the option **Store Selected Query Results** for selected shortcuts in Summary view is also not included.

Compound Queries

Compound Queries make it possible to create complex queries that can combine Node searches with text searches.

1   Go to **Query | Create | New Query → Compound...**
    Default folder is **Queries**.
    Go to 5.

alternatively

1   Click **[Queries]** in Area 1.
2   Select the **Queries** folder in Area 2 or its subfolder.
3   Go to **Query | Create | New Query → Compound...**
    Go to 5.

alternatively

3   Click at an empty space in Area 3.
4   Right-click and select **New Query → Compound...**

The **Compound Query** dialog box is shown. The query is divided into *Subquery 1* and *Subquery 2*. The operator[7] between them can be chosen among several options.

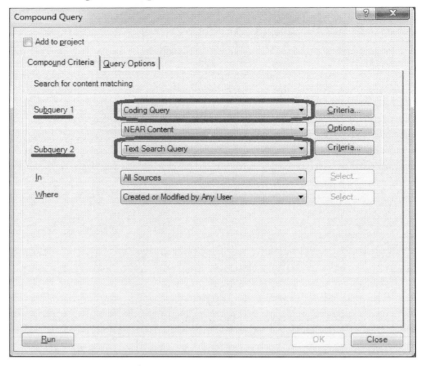

5   Choose *Coding Query* at **Subquery 1**.
6   The **[Criteria...]** button opens the **Subquery Properties** dialog box that is similar the **Coding Query** dialog box only that the option *Add To Project* and the **Query Options** tab are missing.

---

[7] See page 215 onward for explanations of the other operators on this drop-down list.

7. We use the **Advanced** tab and use the following criteria: The Node *Foreign countries* AND the *Age Group 20-29*. See the section about Coding Queries, page 187.
8. Click [**OK**].
9. In the **Compound Query** dialog box select the operator *NEAR Content* and with the [**Options...**] button you select *Overlapping*.
10. Choose *Text Search Query* at **Subquery 2**.
11. The [**Criteria...**] button opens the **Subquery Properties** dialog box, which is similar to the **Text Search Query** dialog box only the option *Add To Project* and the **Query Options** tab missing.

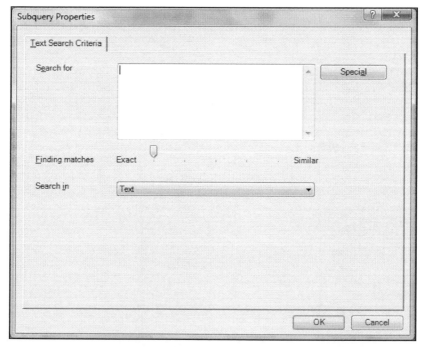

12 Type `excite` in the text box **Search for**, and pull the slider below two steps to the right which includes *Stemmed search* and synonyms.

The **Finding matches** slider is the same as described under **Text Search Queries** (see page 183).

13 Click **[OK]**.

14 Click **[Run]** in the **Compound Query** dialog box.

The format of the results depends on the settings made under the **Query Options** tab in the dialog box **Compound Query** (see page 211).

When you have run a Compound Query with *Preview Only* the result looks like under Text Search Queries, page 184 and onwards, with two exceptions: The Word Tree tab is not included and the option **Store Selected Query Results** for selected shortcuts in Summary view is also not included

Matrix Coding Queries

Matrix Coding Queries have been introduced to display how a set of Nodes relates to another set of Nodes. The results are presented in the form of a matrix or table.

Example: We want to explore how different age groups relate to certain selected themes represented by theme nodes.

1. Go to **Query** | **Create** | **New Query** → **Matrix Coding...**
   Default folder is **Queries**.
   Go to 5.

alternatively
1. Click **[Queries]** in Area 1.
2. Select the **Queries** folder in Area 2 or its subfolder.
3. Go to **Query** | **Create** | **New Query** → **Matrix Coding...**
   Go to 5.

alternatively
3. Click on an empty space in Area 3.
4. Right-click and select **New Query** → **Matrix Coding...**

The **Matrix Coding Query** dialog box appears:

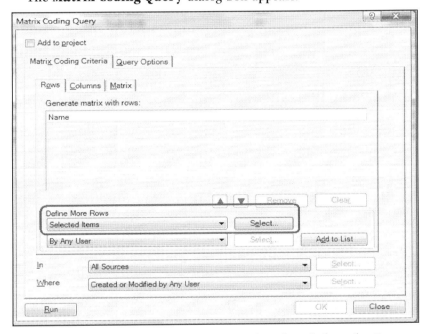

5. Select the **Matrix Coding Criteria** tab and then the **Rows** tab.
6. Choose *Selected Items* from the **Define More Rows** drop-down list.
7. Click **[Select...]**.

The **Select Project Item** dialog box appears:

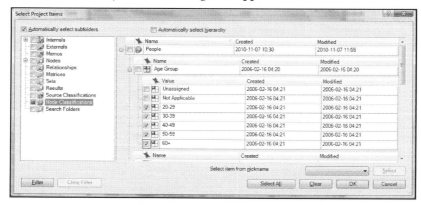

8   Select **Node Classifications\\People\Age Group** and check the values that you want to use. Click **[OK]**.
9   Click **[Add to List]**.

The result may look like this:

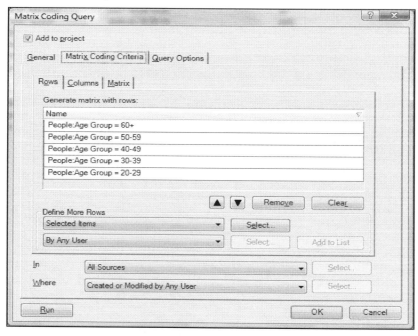

10 Select the **Matrix Coding Criteria** tab and then the **Columns** tab.
11 Choose *Selected Items* from the **Define More Columns** drop-down list.
12 Click **[Select...]**.

198

The **Select Project Items** dialog box appears:

13 Select **Nodes\\Experience** and check the Nodes that you want to study. Click **[OK]**.
14 Click **[Add To List]**.

When you have defined the columns the result may be like this:

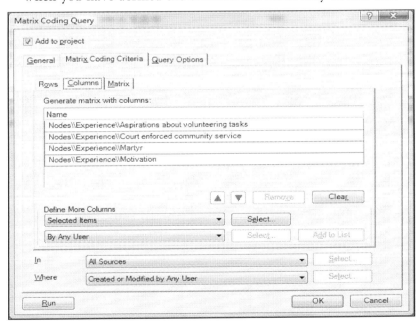

If there are Nodes that you would like to delete, select them and click **[Remove]**. The whole list is cleared with **[Clear]**. If you want to change the order then select a Node and use the arrow buttons to move up or down.

15 Select the **Matrix Coding Criteria** tab and then the **Matrix** tab.

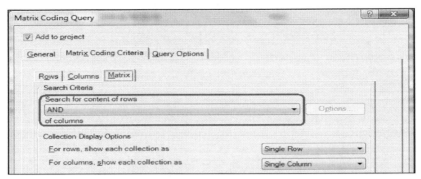

You can now choose operator[8] to use between rows and columns.

16 Click [**Run**] in the **Matrix Coding Query** dialog box.

The format of the result depends on the settings under the **Query Options** tab in the dialog box **Matrix Coding Query** (see page 211).

The option *Preview Only* displays the matrix in Area 4 and may look like this:

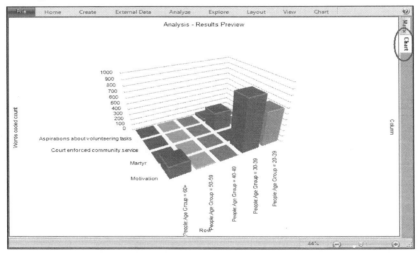

You can also show a Chart of this matrix. Click the *Chart* tab on the right side of the window:

---

[8] See page 213 onward for explanations of the other operators on this drop-down list.

The ribbon menu **Chart** opens when the matrix is showing as a Chart. The Chart options allow adjusting formatting, zooming and rotating. By going to **Chart | Type** the following drop-down menu appears:

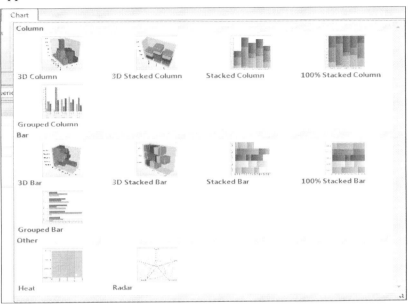

Here you can choose from various types of diagrams.

Opening a Cell

A matrix is a collection of cells. Each cell is a Node. You may need therefore to study each cell separately.

1. Open the matrix.
2. Select the cell you want to open.
3. Right-click and select **Open Matrix Cell**
   or double-click the cell.

The cell opens and can be analyzed as any other cell. This Node is an integral part of the matrix and if you want to save it as a new Node then select the whole Node in the *Reference* view mode and go to **Analyze | Coding | Code Selection At → New Node** or right-click and select **Code Selection → Code Selections At New Node** or **[Ctrl] + [F3]**.

Viewing Cell Content

There are several options to view cell content when cells are not opened.

1. Open the matrix.

2 Go to **View | Detail View | Matrix Cell Content** → <select>
or right-click and select **Matrix Cell Content** → <select>
any of the following options:

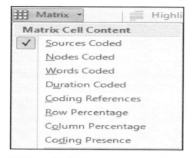

Hiding/Unhiding Row Numbers
1 Open the matrix.
2 Go to **Layout | Show/Hide | Row IDs**
or right-click and select **Row** → **Row Ids**.

Hiding Rows
1 Open the matrix.
2 Select one or more rows that you want to hide.
3 Go to **Layout | Show/Hide | Hide Row**
or right-click and select **Row** → **Hide Row**.

Hiding/Unhiding Rows with Filters
1 Open the matrix.
2 Click the 'funnel' in a certain column head
or select a column and go to **Layout | Sort & Filter | Filter** → **Filter Row**.

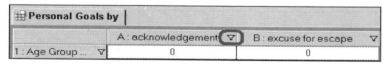

The **Matrix Filter Options** dialog box appears:

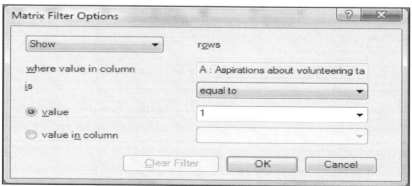

202

3   Select value and operator for hiding or unhiding. Confirm with [**OK**]. When a filter is applied the funnel turns *red*.

To clear a filter use [**Clear Filter**] in the **Matrix Filter Options** dialog box.

Unhiding Rows
1. Open the matrix.
2. Select one row on each side of the hidden row that you want to unhide.
3. Go to **Layout | Show/Hide | Unhide Row**
   or right-click and select **Row → Unhide Row**.

Unhiding All Rows
1. Open the matrix.
2. Go to **Layout | Sort & Filter | Filter → Clear All Row Filters**
   or right-click and select **Row → Clear All Row Filters**.

Hiding/Unhiding Column Letters
1. Open the matrix.
2. Go to **Layout | Show/Hide | Column IDs**
   or right-click and select **Column → Column IDs**.

Hiding Columns
1. Open the matrix.
2. Select one or more columns that you want to hide.
3. Go to **Layout | Show/Hide | Hide Column**
   or right-click and select **Column → Hide Column**.

Unhiding Columns
1. Open the matrix.
2. Select one column on each side of the hidden column that you want to unhide.
3. Go to **Layout | Show/Hide | Unhide Column**
   or right-click and select **Column → Unhide Column**.

Unhiding All Columns
1. Open the matrix.
2. Go to **Layout | Sort & Filter | Filter → Clear All Column Filters**
   or right-click and select **Column → Clear All Column Filters**.

Transposing the Matrix
Transposing means that rows and columns are changing places.
1. Open the matrix.
2. Go to **Layout | Transpose**
   or right-click and select **Transpose.**

Moving a Column Left or Right
1. Open the matrix.
2. Select the column or columns that you want to move. If you want to move more than one column they need to be adjacent.
3. Go to **Layout | Rows & Columns | Column → Move Left/Move Right**.

Resetting the Whole Matrix
1. Open the matrix.
2. Go to **Layout | Tools | Reset Settings**
   or right-click and select **Reset Settings**.

Viewing the Cells Shaded or Colored
1. Open the matrix.
2. Go to **View | Detail View | Matrix → Matrix Cell Shading → <select>**
   or right-click and select **Matrix Cell Shading → <select>**.

Exporting a Matrix
1. Open or select the matrix.
2. Go to **External Data | Export | Export Matrix...**
   or right-click and select **Export Matrix...**
   or **[Ctrl] + [Shift] + [E]**.

The **Save As** dialog box is shown and you can decide the file name and file location and create a text file or an Excel spreadsheet.

When you view a Chart you can export the image in the following formats: .JPG, .BMP or .GIF.

Converting a Matrix to Nodes
There are situations when you need to convert cells in a matrix to Nodes.
1. Open or select the matrix.
2. Copy by going to **Home | Clipboard | Copy**
   or right-click and select **Copy**
   or **[Ctrl] + [C]**.
3. Click **[Nodes]** in Area 1.
4. Select the **Nodes** folder in Area 2 or its subfolder.
5. Go to **Home | Clipboard | Paste → Paste**
   or right-click and select **Paste**
   or **[Ctrl] + [V]**.

The **Paste** dialog box appears:

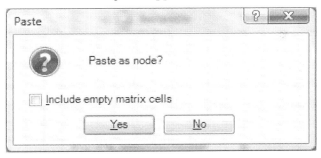

6   Confirm with [**Yes**].

The result is a hierarchical Node[9] where the Parent Node inherits the name of the matrix, called 'Matrix Parent'. The first generation Child Nodes are the rows, called 'Row Parents' and the grandchildren Nodes contain contents from each cell.

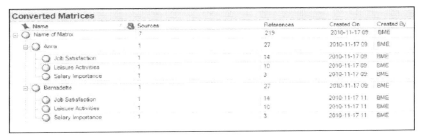

These Nodes can then be used for Cluster Analysis (see page 307).

---

[9] When *converting a Node Matrix* to Nodes then the "Matrix Parent" and the "Row Parents" are calculated with the *Aggregate* function. This method is giving the correct number of Sources but the number of references is suffering from the imperfection we mention on page 129.

Group Queries

Use Group Queries to find items that are associated in a particular way with other items in your project. You could for example explore the difference in coding between sources (scope items) with a Group query. When you run the query, the results are displayed in Detail View with the coded Nodes grouped under each scope item.

1. Go to **Query | Create | New Query → Group...**
   Default folder is **Queries**
   Go to 5.

alternatively

1. Click on **[Queries]** in Area 1.
2. Select the **Queries** folder in Area 2 or its subfolder
3. Go to **Query | Create | New Query → Group...**
   Go to 5.

alternatively

3. Click on any empty space in Area 3.
4. Right-click and select **New Query → Group...**

In each case, the **Group Query** dialog box is shown:

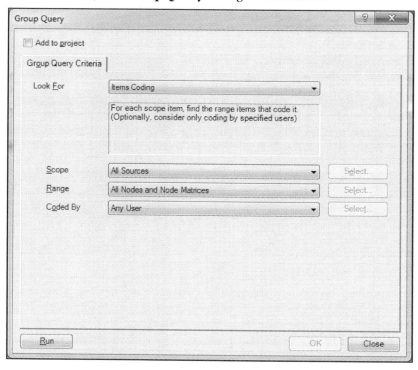

The **Look For:** drop-down list has the following options:

Items Coding	For each scope item, find the range items that code it. (Optionally, consider only coding by specific users)
Items Coded At	For each scope item, find the range items that it codes. (Optionally, consider only coding by specific users)
Items by Attribute Value	For each attribute value in the scope, find the items in the range that have that value assigned.
Relationships	For each scope item, find the items that it has a relationship of the selected direction/type with.
See Also Links	For each scope item, find the range items that it has s See Also link with.
Model Items	For each model in the scope, find the range items that appear in the model.
Models	For each scope item, find the models in the range that it appears in.

Depending on what options you select the **Scope** and **Range** drop-down lists offers corresponding alternatives.

Let's assume that you need to explore what Nodes two selected sources are coded at.

5   Select *Items Coding* from the **Look For** drop-down list.
6   Select *Selected Items* from the **Scope** drop-down list.
7   Click the **[Select]** button and from the **Select Project Items** dialog box select for example two Source Items (two interviews). Click **[OK]**.
8   Select *Selected Folders* from the **Range** drop-down list.
9   Click the **[Select]** button and from the **Select Folders** dialog box select for example the **Theme Nodes** folder. Click **[OK]**.
10 Finally click **[Run]** in the **Group Query** dialog box.

Group Query Results are displayed as an expandable list in Area 3. This list cannot be saved. The query, however, can be saved as any other queries, see Chapter 14, Common Query Features. When the saved query is run the expandable list appears again.

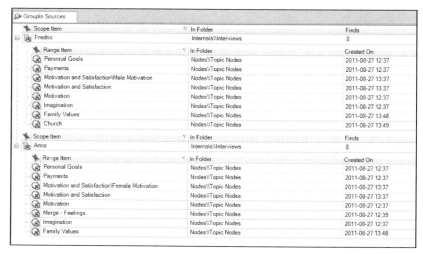

Selecting the **Connection Map** tab to the right the following graph is shown:

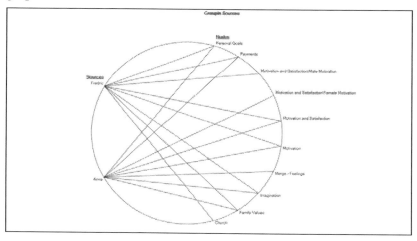

In a corresponding way you can select any scope item like Node(s), model(s), and then the range of its related items.

14. COMMON QUERY FEATURES

This chapter deals with the functions and features common to several types of queries. The Filter function which is described here is an example of a common query feature. One way to benefit from the filter function is letting the filter eliminate unwanted items. For example, you can use the filter to eliminate Nodes that were created later than last week.

The Filter Function

The [**Select...**] button is available in many dialog boxes when queries are created. This button always opens the **Select Project Items** dialog box:

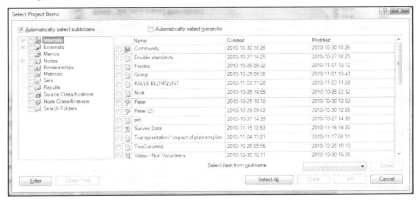

Automatically select subfolders means that when a folder is selected in the left hand window all the underlying subfolders and items will be selected. Folders which cannot have subfolders (Nodes, Sets, and Results) will select all of the items therein.

Automatically select hierarchy means that when a certain item in the right hand window has been selected all underlying items are also selected.

The [**Filter**] button is always available at the bottom left corner of the **Select Project Items** dialog box and this button opens the **Advanced Find** dialog box:

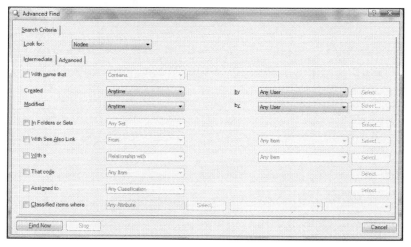

These are the same as the Advanced Find search functions (see page 280).

Saving a Query

As mentioned at the start of previous chapter, Queries made can be saved so that they can be run again at a later stage. Let's make an example with a Text Search Query where we have already completed the Text Search Criteria.

1   In the **Text Search Query** dialog box check *Add to project*, and a new **General** tab will show:

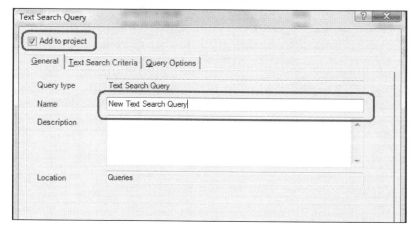

2   Type a name (compulsory) and a description (optional), then click [**Run**] or [**Apply**] or [**OK**].

210

Saving a Result

The result of a query can be displayed on the screen using the option *Preview Only*. The result is shown in Area 4 but not saved.

Preview Only for Text Search Queries opens the Summary tab (see page 162).

Preview Only for Coding Queries and Compound Queries open the Reference tab (see page 162).

Preview Only for Matrix Search Queries opens the Node Matrix tab (see page 200).

If you want to save the result as a Node there are a few options to choose from, for example *Create Results as New Node*.

1. In the **Text Search Query** dialog box select the **Query Options** tab.

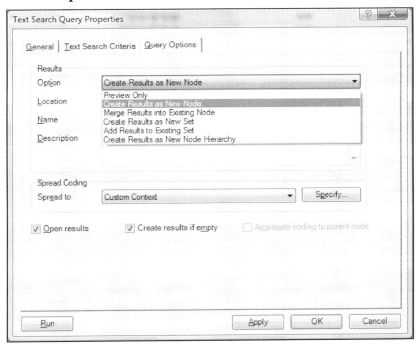

2. Choose *Create Results as New Node* from the **Results/Options** drop-down list. Check *Open results* when you want open the Node when the query is run. Check *Create results if empty* if you want to create an 'empty' Node when zero result.

3   Accept default **Location** *Results*[10] or use **[Select]** and choose another location, for example *Nodes* when you want to use this Node for future coding or else future editing.
4   Type a name (compulsory) and a description (optional), then **[Run]**.

The **Spread Coding** drop-down list allows you to decide options for coding surround sections of data nearby your coded query result. The following options[11] are available:
- None
- Coding Reference (applicable when searching in Nodes)
- Narrow Context
- Broad Context
- Custom Context
- Entire Source

- ♦ -

Using **Query | Create | Last Run Query** the <...> **Query Properties** dialog box is shown again and you may modify or edit the query. Each time a query is run it is also saved provided such option has been selected. When editing a query you can for example apply Surrounding Paragraphs at the **Spread Coding** drop-down list.

About the Results Folder

The Results folder is the default folder where a result of query is saved. You can however modify **Query Properties** so that query results will be saved in any Node location. But there are some advantages to using the Results folder.

First, it is practical to see if the result is reasonable (before it is saved in its final location or made into a Node) or if the query needs immediate modification. Sometimes, when the query is not saved but the result is, the command **Open Linked Query**, allows you to to open or modify the query.

Nodes in the Results folder cannot be edited or used for further coding or uncoding and commands like **Uncode At this Node** and **Spread Coding** are unavailable. After verifying your Node in the Results folder you should move the result to a location under Nodes, where it can be more fully analyzed.

---

[10] Storing results in the Results folder means that the node cannot be edited nor can it be used for further coding or uncoding.

[11] The definition of *Narrow* and *Broad* is determined under **Application Options**, the **General** tab, see page 36. *Custom* can override these settings for any specific task.

When you run a Text Search Query that is saved in the Results folder Coding Context Narrow (5 words) is activated, but the Coding Context is reset as soon as the Node is moved to a location under Nodes. If you then should need Coding Context this feature can be activated with a separate command (see page 165).

Editing a Query

A saved query can be run anytime:
1. Click **[Queries]** in Area 1.
2. Select the **Queries** folder in Area 2 or its subfolder.
3. Select the query in Area 3 that you want to run.
4. Go to **Query | Create | Run Query**
   or right-click and select **Run Query...**

You can always optimize a saved query so that it fulfils your changing needs. Or you may wish to copy a query before editing:
1. Click **[Queries]** in Area 1.
2. Select the **Queries** folder in Area 2 or its subfolder.
3. Select the query in Area 3 that you want to edit.
4. Go to **Home | Item | Properties**
   or right-click and select **Query Properties...**
   or **[Ctrl] + [Shift] + [P]**.

One of the following dialog boxes is shown:
- **Text Search Query Properties**
- **Word Frequency Query Properties**
- **Coding Query Properties**
- **Matrix Coding Query Properties**
- **Compound Query Properties**
- **Coding Comparison Query Properties**

The **Coding Query Properties** dialog box, for example, has the same contents as the **Coding Query** dialog box. Here you can make your modifications.

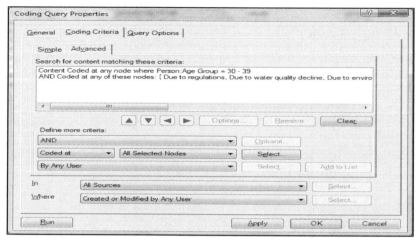

[**OK**] will carry out the modifications without running the query.

[**Apply**] will carry out the modifications without running the query again and the dialog box will remain shown so that more modifications can be made.

[**Run**] will carry through the modifications and run the query. If the option Create Results as a New Node under the **Query Options** tab has been selected then another Node will be created in the given folder. If you choose to let the results initially be located under the **Results** folder they can be moved later on.

The Operators

In the **Coding Query, Matrix Coding Query and Subquery Properties** dialog boxes there are drop-down lists with various operators: AND, OR, NEAR, PRECEDING and SURROUNDING. The following charts explain the results when these operators are applied.

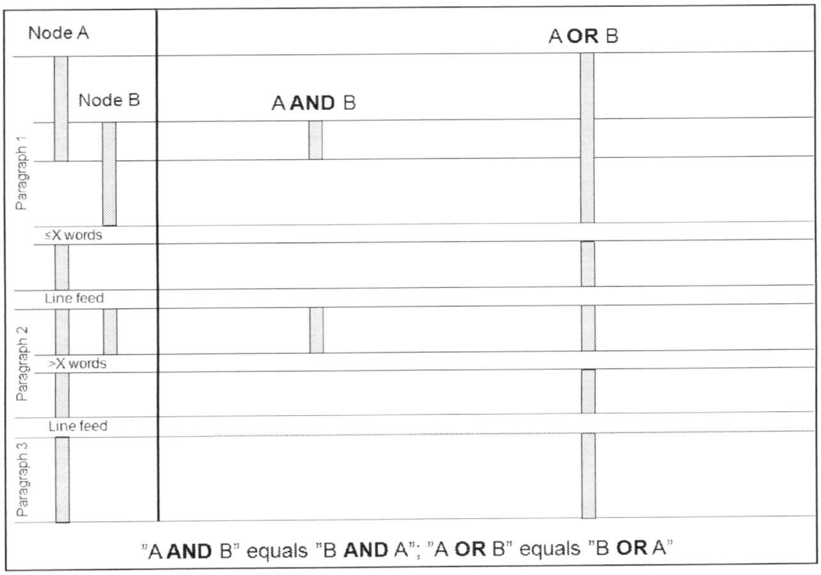

Node A
A **OR** B
Node B
A **AND** B
Paragraph 1
≤X words
Line feed
Paragraph 2
>X words
Line feed
Paragraph 3
"A **AND** B" equals "B **AND** A"; "A **OR** B" equals "B **OR** A"

AND displays the elements of a document where both A and B have been coded.

OR displays the elements of a document where either A or B or both A and B have been coded.

AND NOT displays the elements of a document where A but not B have been coded.

NEAR Content Overlapping displays the elements of a document where A and B are overlapping.

NEAR Content In Custom Context[12]. The **[Specify]** button allows you to select between Broad Context, Narrow Context or Custom Context.

For example:
- **NEAR Content Within X words** displays the elements of a document where A and B are within X words from each other.
- **NEAR Content In Surrounding Paragraph** displays the elements of a document where A and B are within same paragraph delimited by line feed.

NEAR Content In Same Scope Item displays the elements of a document where A and B are within the same document.

NEAR Content In Same Coding Reference displays the elements of a document where A and B are within the same Node.

---

[12] The definition of *Narrow* and *Broad* is determined under **Application Options**, the **General** tab, see page 36. *Custom* can override these settings for any specific task.

PRECEDING Context Overlapping displays the elements of a document where A and B overlap as long as A is coded earlier or from the same starting point as B.

PRECEDING Content In Custom Context[13]. The **[Specify]** button allows you to select between Broad Context, Narrow Context or Custom Context.

For Example:
- **PRECEDING Context Within X words** displays the elements of a document where A and B are within X words as long as A is coded earlier or from the same starting point as B.
- **PRECEDING Context In Surrounding Paragraph** displays the elements of a document where A and B are within the same paragraph delimited by line feed as long as A is coded earlier or from the same starting point as B.

PRECEDING Context In Same Scope Item displays the elements of a document where A and B are within same document as long as A is coded earlier or from the same starting point as B.

PRECEDING Context In Same Coding Reference displays the elements of a document where A and B are within same Node as long as A is coded earlier or from the same starting point as B.

---

[13] The definition of *Narrow* and *Broad* is determined under **Application Options**, the **General** tab, see page 36. *Custom* can override these settings for any specific task.

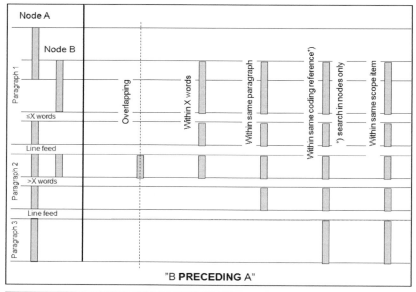
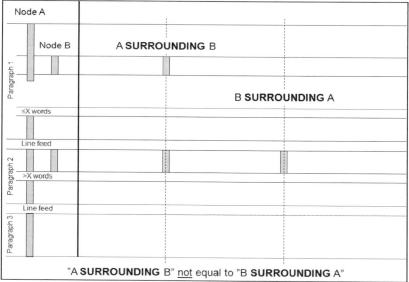

SURROUNDING Context displays the elements of a document where A overlaps B as long as A is coded earlier or from the same starting point as B and terminates later than or at the same point as B.

15. HANDLING BIBLIOGRAPHIC DATA

Along with source material that is gathered as project evidence, reference material (e.g., peer-reviewed academic research papers) often play a crucial role in grounding a qualitative research project. NVivo 10 also allows users to import reference material, including full-text documents, from common reference handling software like EndNote, RefWorks and Zotero. When imported, reference materials become Source Items and as a result they can be coded and analyzed the same way as other sources. For advanced analysis, we offer a method of using the Framework Matrices (see Chapter 16, About the Framework Method) to efficiently work with academic reference material, such as Literature Reviews. This chapter is about importing bibliographic data stored in certain selected reference handling software. The file formats that can be imported to NVivo are: .XML for EndNote and .RIS for RefWorks and Zotero .

In this chapter, we will use as an example importing data into NVivo from EndNote (the top reference handling software, in our opinion). The following two reference records will be exported from Endnote:

	Author	Year	Title	Journal	Ref Type	URL	Last Updated
𝟅	Cafazzo, J...	2009	Patient-perceived barriers to the ado...	Clinical jour...	Journal Arti...	http://www.ncbi.nlm...	2011-08-24
	Ritchie, L.; P...	2011	An exploration of nurses' perceptions...	Applied nur...	Journal Arti...	http://www.ncbi.nlm...	2011-08-24

The clip symbol indicates that one reference has a file attachment (typically a PDF full text article) and the other not.

On the next page, you will see a shot of a typical reference from EndNote. As you can see, each reference listing contains a wealth of meta-data about a single reference:

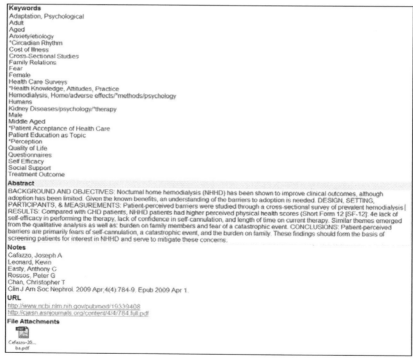

The communication between reference handling software and NVivo is in the form of an XML structured file. Reference handling software typically allows you to export a collection of citations. The export command from for example EndNote is **File → Export** and the file type must be set as XML. This creates you a file with all the above information including a file path to the PDF.

Importing Bibliographic Data

In NVivo go to **External Data | Import | From Other Sources → From EndNote...**. With the file browser you will find the XML-file you exported from your reference handling software. Click **[Open]**.

The **Import from EndNote** dialog box appears:

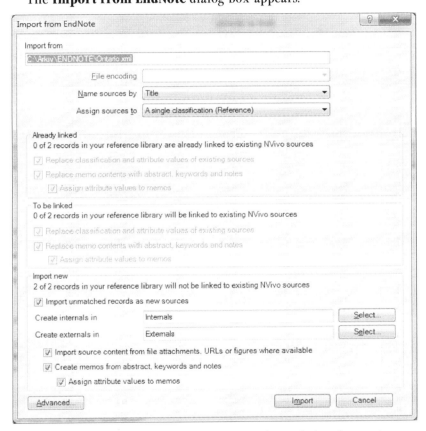

The first option is under **Name sources by** and the alternatives are: Title and Author and year.

The second option is to decide if you want one Source Classification for all your bibliographic data, Reference. Then there will be one attribute called Reference Type and the values will be Journal Article, Book, Conference Proceedings etc. If this is your preference then select

Assign sources to: *A single classification (Reference).*

If you instead prefer one Source Classification for each reference type then select

Assign sources to: *Different classifications based on record type.*

The next option is under the section **Import new** at the bottom of the screen. The first example is a reference with a linked PDF and the second example without PDF. The principle is that PDFs will be imported as Internal Source Items and other references will be imported as External Source Items.

Under the section **Import unmatched records as new sources** you need to define one location for Internal sources and one location for External sources. In our example we have created these two folders:

Internals\\Bibliographic Data and
Externals\\Bibliographic Data.

The option *Import source content from file attachments, URLs or figures where available* is necessary when you want to import a PDF or any other resource. If you uncheck this option then the record will be imported as an External item.

The option *Create memos from abstract, keywords and notes* is selected when each bibliographic item will have a linked memo with the mentioned content.

The option *Assign attribute values to memos* assigns same classification and same attribute values to the linked memo as the linked item.

The [**Advanced**] button makes it possible for some individual settings for the items that is about to be imported. This is useful when you need to import bibliographic data as an update to previously imported data.

The PDF Source Item

The internal PDF Source Item has the same look and layout as the original article and can now be coded, linked, searched and queried:

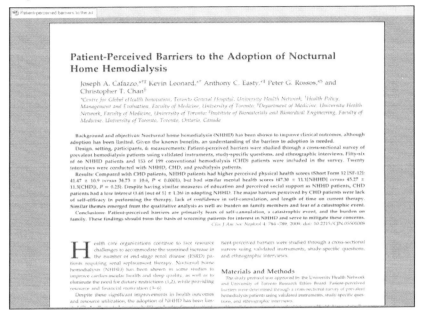

The **PDF Properties**, **General** tab, has now the following content imported through the XML file. As you can see, the abstract has been copied into the Description field of the PDF source:

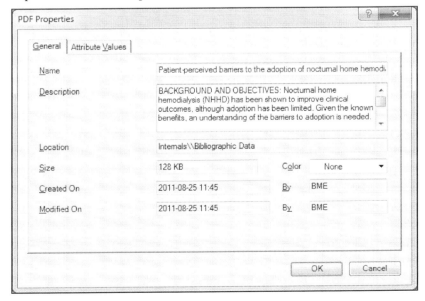

The **PDF Properties**, **Attribute Values** tab, has now the following content:

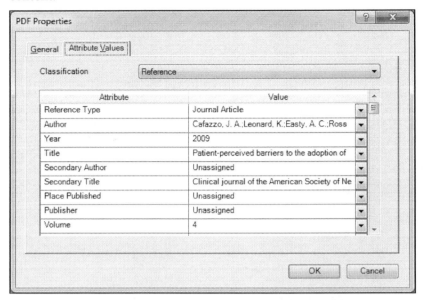

Author is one of the Attributes and the values are the list of author names for each Source Item.

Keywords is also an Attribute (not shown here) and its values are the whole list of keywords originating from this Source Item (see page 220).

The Linked Memo

If you elect to create a linked memo it will have the same name as the linked item. The memo is a normal text document and can be edited and otherwise handled as any Source Item. The content in our example is from the Abstract, Keywords and Notes fields of the original reference record. This is a useful feature because it allows you to search and code the abstract, which is not possible when the abstract is only located in the source description field.

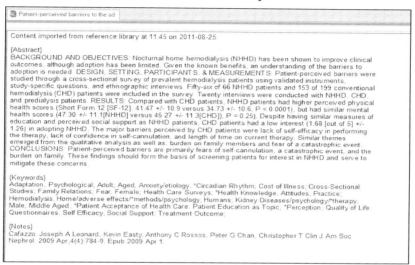

The **Memo Properties**, **General** tab, has now the following content:

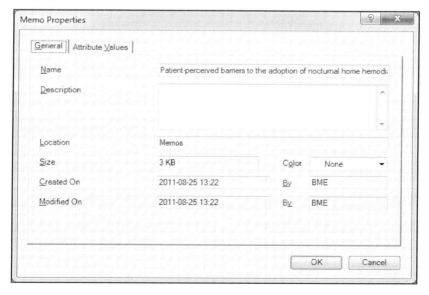

The **Memo Properties**, **Attribute Values** tab, has now the following content:

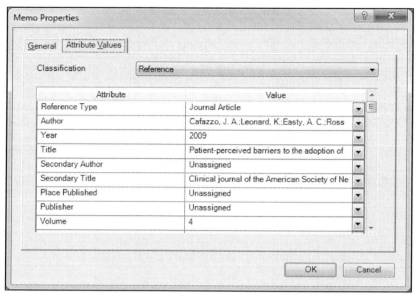

The Classification and Attribute values are identical with those of the linked item.

16. ABOUT THE FRAMEWORK METHOD

Framework is a qualitative data analysis method developed by the UK's largest, independent not-for-profit research institute, the National Centre for Social Research (NatCen) in the 1980's.

The Framework method is used to organize and manage research through the process of summarization, resulting in a robust and flexible matrix output which allows the researcher to analyze data both by case and theme. It's used by hundreds of researchers in areas such as health research, policy development and program evaluation.

NatCen developed specialty software called FrameWork to support this method. This software is no longer developed, but through a partnership between NatCen and QSR, NVivo 10 now provides new functionality to support the Framework method.

Accordingly, the Framework approach will provide you with exciting opportunities to apply this method to textual and non-textual data (audio-visual or images) and adopt other approaches that NVivo also supports such as discourse analysis.

Framework differs from traditional qualitative approaches to analysis as it does not rely on coding and indexing alone.

Introducing the Framework Matrix

Like any matrix, the Framework Matrix consists of rows and columns. Therefore, for those of you who are familiar with Node Matrices (as a result of a Matrix Coding Query) this approach seems familiar. Thus, rows are Nodes, columns are Nodes, and the cell content is the intersection (or cross coding) between two Nodes also understood as the result of an AND operator. The important difference between a Framework Matrix and a Matrix Coding Query is that the cells of a Framework Matrix can display data, or they can display any text you enter along with any links you create.

More than a tool for displaying your data, a Framework Matrix is a Source Item that allows you to quickly and easily view your data and write notes and insights about it. One particularly useful function of a Framework Matrix is viewing your data in the Associated View while recording your insights in cells. A second useful function is the ability to easily view a certain Theme Node in a grid that allows you to quickly compare data to other Theme Nodes or Case Nodes. A third useful function is the ability to create links between cell content and your source material using summary links.

This is the default view of a Framework Matrix:

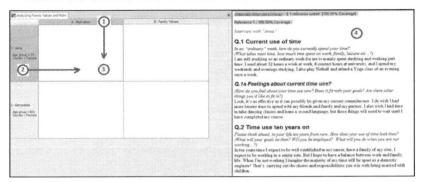

① Rows are defined as Nodes which are often classified Case Nodes. This could be a person or a place or an organization. It could also be literature of any kind for example a set of articles in PDF format. In the latter case PDF Source Items must be created as Nodes before they can be used as rows in a Framework Matrix.
Go to **Create | Items | Create As → Create As Node(s)**.

② Columns are defined as Nodes, typically topic or theme Nodes. It could also be Nodes representing interview questions from structured interviews.

③ The content of a cell is always blank (default) when the first Framework Matrix in a project is created. Several options are now at hand for the user:
You can type any text.
You can let the cell contain the whole or part of the coded content in the intersection between a row and a column.
See Auto Summary, see page 231.
You can create Summary links to any content in the Associated view. See Summary Links, see page 232.

④ The Associated View is a separate window to the right of the Framework Matrix showing the whole or parts of the Case Node of the selected cell or row. There are several options for the Associated View: The whole Node (Row coding), the intersection between a row and a column (Cell coding) or Summary links.

Creating a Framework Matrix

1. Go to **Create | Sources | Framework Matrix**.
   Default folder is **Framework Matrices**.
   Go to 5.

alternatively

1. Click on **[Sources]** in Area 1.
2. Select the **Framework Matrices** folder in Area 2 or its subfolder.
3. Go to **Create | Sources | Framework Matrix**.
   Go to 5.

alternatively

3. Click on any empty space in Area 3.
4. Right-click and select **New Framework Matrix...**
   or **[Ctrl] + [Shift] + [N]**.

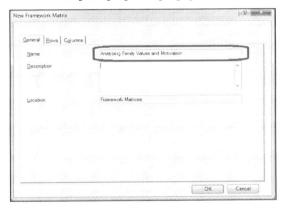

4. In **New Framework Matrix** dialog box type a name (compulsory) and a description (optionally).
5. Click on the **Rows** tab.

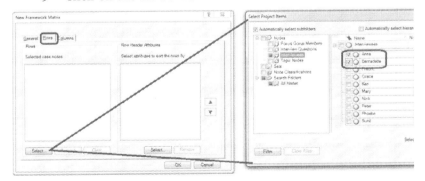

6. Click on the left **[Select]** button and in the **Select Project Items** dialog box select the Nodes (Case Nodes) that you want to become rows in the Framework Matrix, then confirm with **[OK]**.

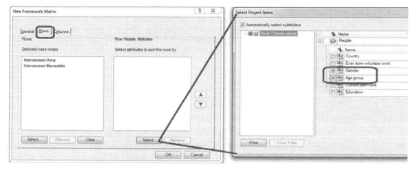

7   Click on the right **[Select]** button and in the **Select Project Items** dialog box select the attributes that you want to become Row Header Attributes in the Framework Matrix, then confirm with **[OK]**.

8   Click on the **Columns** tab.

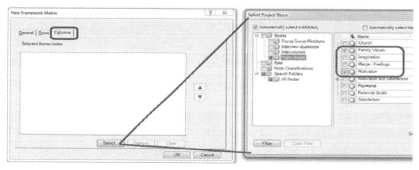

9   Click on the **[Select]** button and in the **Select Project Items** dialog box select the Theme Nodes that you want to become columns in the Framework Matrix, then confirm with **[OK]**.

10 When you are finished choosing your Rows and Columns, click **[OK]**.

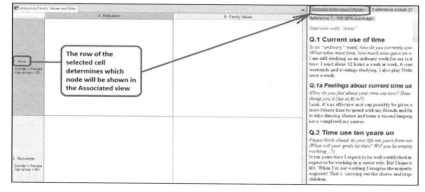

When your Framework Matrix is finished, you will see a blank screen resembling the above image. The default cell content is blank.

Remember, a Framework Matrix is a Source Item, so you can create memo-like insights and their attendant links as a way of writing up your insights or generating qualitative data.

Populating Cell Content

With your new Framework Matrix created, you have three options for populating content in the cells:

- auto-populating the cell with all or part of the content at the 'intersection' between a row and a column. (See Auto Summary below)
- creating Summary links to any content in the Associated view. (See Summary Links, page 232)
- typing in any text you wish

Auto Summary

1 Go to **Analyze | Framework Matrix | Auto Summary**.

Using Auto Summary, all cells (irrespective of which cell is selected) will be automatically filled with the content corresponding to the intersection between a row and a column.

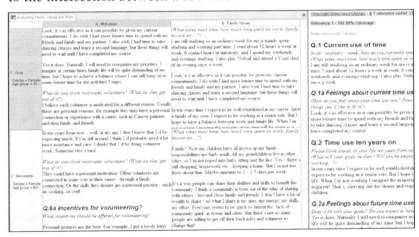

If there is any text in a cell before Auto Summary is applied, then the new content will be pasted after that text. Using Auto Summary will not overwrite extant cell content. Using Auto Summary repeatedly will create repeated content in all cells.

After an Auto Summary, you can modify text in any way you see fit. One best practice we recommend is populating cells with Auto-Summary and then writing your own summary overtop of the content. In this way, you can easily record your insights on aspects of your data.

Summary Links

Summary links are connection points between Framework Matrix content and content from your data sources. Like See Also Links, Summary Links allow for shortcutting across your project and moving seamlessly between summary content and your data.

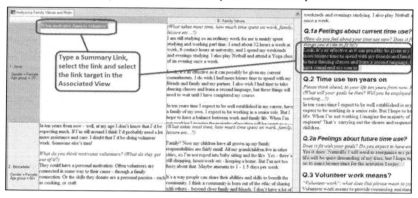

1. If required blank out the current cell content with **[Ctrl]** + **[A]** then the **[Del]** key.
2. Type the text of the new Summary Link.
3. Select this link.
4. Select the linked content in the Associated View.
5. Go to **Analyze | Framework Matrix | New Summary Link**.

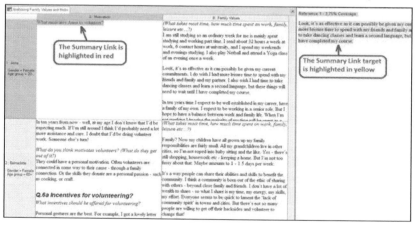

To get the above view you need the following settings:
Go to **View | Detail View | Framework Matrix**:
→ **Summary Links | Show**
→ **Associated View Content | Summary Links**
→ **Associated View Highlight | Summary Links**

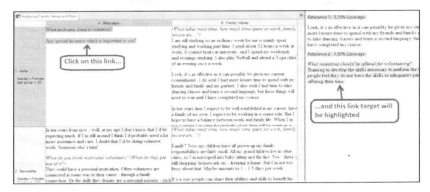

If you need more than one Summary Link in a cell then it is very handy to use the setting:

Go to **View | Detail View | Framework Matrix**:
→ **Summary Links | Show**
→ **Associated View Content | Summary Links**
→ **Associated View Highlight | Summary Links from Position**

More on Associated View

The default settings for the ribbon **View | Detail View | Framework Matrix** is determined in the **Application Options** dialog box. Go to **File → Options → Display** tab, section **Framework Matrix Associated View Defaults**:

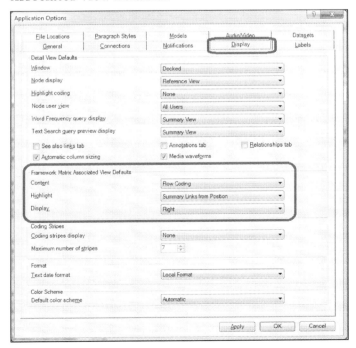

This is the result of the above default settings under **View | Detail View | Framework Matrix**:

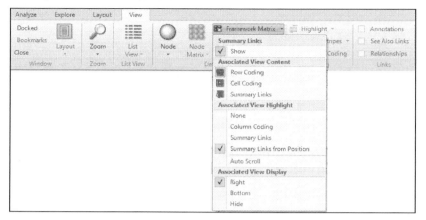

These default settings are restored each time a project is opened and any changes made are kept intact during the current work session.

1. Go to **View | Detail View | Framework Matrix**
   → **Associated View Highlight | Column Coding**.

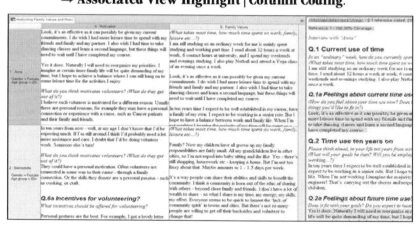

The coded sections are now highlighted depending on which cell has been selected.

1   Go to **View | Detail View | Framework Matrix**
    → **Associated View Content | Cell Coding**.

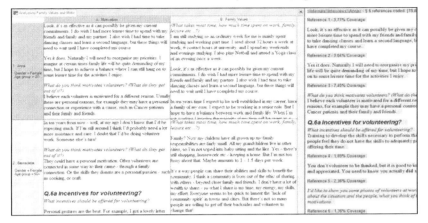

The Associated View now shows only the coded content depending on which cell has been selected.

- ◆ -

The Associated View can be displayed to the right of or below the Framework Matrix:
Go to **View | Detail View | Framework Matrix**
→ **Associated View Display | Right** or **Bottom**
or hidden:
Go to **View | Detail View | Framework Matrix**
→ **Associated View Display | Hide**

Working with Framework Matrices

Auto Scroll
Auto Scroll scrolls the Associated View as follows. When you click on a Summary link in a cell then the Associated View displays the currently highlighted section. If Highlight Column Coding is chosen then the first highlighted section is the coded section and if Highlight Summary Links is chosen then the first highlighted section is the Summary Link. The option Summary Links from Position is useful when you have created more than one Summary Link in a cell.

Where is the Cell Summary stored?
The cell Summary is stored in the intersection between two Nodes, one from a row and one from a column in a Framework Matrix. Once created the Summary is stored even if the Framework is deleted. If same combination of two Nodes occurs in another Framework the Summary is identical. Changing the Summary in one Framework is therefore instantly mirrored in the other Framework.

Opening a Framework Matrix
1. Click on [**Sources**] in Area 1.
2. Select the **Framework Matrices** folder in Area 2 or its subfolder.
3. Select the Framework Matrix in Area 3 that you want to open.
4. Go to **Home | Item | Open**
   or right-click and select **Open Framework Matrix...**
   or double-click on the Framework Matrix in Area 3
   or [**Ctrl**] + [**Shift**] + [**O**].

Please note, you can only open one Framework Matrix at a time, but several matrices can stay open simultaneously.

Editing a Framework Matrix
1. Select a Framework Matrix.
2. Right-click and select **Framework Matrix Properties...**

You can add or delete rows and columns.

Importing Framework Matrices
Framework Matrices can be imported along with another project that you import. All Nodes which constitute the Framework Matrix must either exist in the open project or must be imported with the Framework Matrix. The Framework Matrix will be updated with the updated Nodes.

Exporting Framework Matrices
1. Click **[Sources]** in Area 1.
2. Select the **Framework Matrices** folder in Area 2 or its subfolder.
3. Select the Framework Matrix or Matrices in Area 3 that you want to export.
4. Go to **External Data | Export | Export → Export Framework Matrix...**
   or right-click and select **Export → Export Framework Matrix...**
   or **[Ctrl]** + **[Shift]** + **[E]**.
5. Decide file name, file location, and file type. Possible file types are: .TXT, .XLS, or XLSX. Confirm with **[Save]**.

Deleting a Framework Matrix
1. Click on **[Sources]** in Area 1.
2. Select the **Framework Matrices** folder in Area 2 or its subfolder.
3. Select the Framework Matrix or Matrices in Area 3 that you want to delete.
4. Go to **Home | Editing | Delete**
   or right-click and select **Delete**
   or **[Del]**.
5. Confirm with **[Yes]**.

Please note, that according to what was said about storing Framework Matrices the content of a matrix is not deleted even when the matrix is. The content is saved as the intersection between two Nodes. When one of those Nodes is deleted the content will be deleted.

Printing Framework Matrices
1. Open a **Framework Matrix**.
2. Go to **File → Print → Print...**
   or right-click and select **Print...**
   or **[Ctrl]** + **[P]**.

Undocking the Framework Matrix
Undocking a Framework Matrix is possible with the same command as for other open Project Items. However, the Associated View window is hidden in this view. One way to use more screen space and still keep the Associated view is closing the Navigation View by going to **View | Workspace | Navigation View**.

Fonts, Font Styles, Size, and Color

The default text style in the Cell Summary is determined by **Project Properties** dialog box, the **Framework Matrices** tab, see page 58:

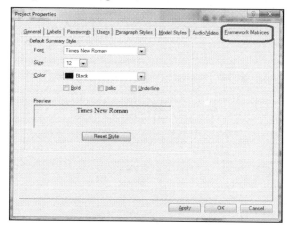

You can add text styles and attributes as an overlay to these default settings:
1. Select the text in a cell.
2. Go to **Home | Format |** and select font, size, color or attribute.

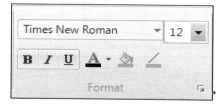

Setting paragraph styles under **Home | Styles** is not available for Framework matrices.

Paragraph alignment, indentation, bulleted or numbered lists under **Home | Paragraph** is not available for Framework Matrices.

Searching and Replacing Words

This feature is the same as when editing Source Items that are documents.
 Home | Editing | Find → Find...
 Home | Editing | Replace
 However, **Home | Editing | Find → Go to...** is not available for Framework Matrices.

Spell Checking

NVivos native spell checking can be used for Framework Matrices, see page 74.

Inserting Date and Time and Symbols
This feature is the same as when editing Source Items that are documents.
    Home | Editing | Insert → Date/Time
    Home | Editing | Insert → Symbol...

Organizing Framework Matrices

About Sorting Rows in a Framework Matrix
Rows in a Framework Matrix are sorted according to the selected Attributes and attribute values that are displayed under the name of each row. If no Attributes have been chosen for display the Node names of the rows are sorted alphabetically.

About Sorting Columns in a Framework Matrix
Columns in a Framework Matrix are sorted according the setting in the Framework Matrix Properties dialog box where you can change the sort order. You can also select a column, right-click and select **Column → Move Left** ([Ctrl] + [Shift] + [L]) or **Column → Move Right** ([Ctrl] + [Shift] + [R]). Alternatively, select a column go to **Layout | Rows & Columns | Column → Move Left** or **Move Right**.

Hiding and Filtering Rows and Columns
Hiding and filtering rows and columns in a Framework Matrix (like you could for a matrix Node) is not possible in a Framework Matrix.

Adjusting Row Height and Column Width
    1   Select a row or a column.
    2   Right-click and select **Row → Row Height (Column Width)**.
    3   Enter the row height (column width) in pixels.
    4   Confirm with **[OK]**.

The row height and column width can also be adjusted by pointing at the border between rows or between columns and drag with the mouse pointer.

Row heights can also be set to automatically adjust to the amount of text, however maximum value for autofit is 482 pixels:
    1   Select a row or rows.
    2   Right-click and select **Row → AutoFit Row Height**.

If you need more row height, then use **Row → Row Height** instead.

Row heights and Column widths can be reset to default values (values applied when you create a new Framework Matrix).
    1   Select any row, column or cell in the Framework Matrix,
    2   Go to **Layout | Tools | Reset Settings**
        or right-click and select **Reset Settings**.

17. ABOUT QUESTIONNAIRES AND DATASETS

This section deals with data that originates from both multiple-choice questions and open-ended questions. In NVivo a dataset is a Source Item in NVivo created when structured data is imported. Structured data is organized in records (rows) and fields (columns). The structured data formats that NVivo can import are Excel spreadsheets, tab-delimited text files and database-tables compatible with Microsoft's Access. A dataset in NVivo is presented in a built-in reader that can display data in both a table format and in a form format. The reader makes it much easier to work on the computer and read and analyze data.

A dataset has two types of fields (columns), namely Classifying and Codable.

Classifying is a field with demographic content of a quantitative nature, often the result of multiple choice questions. The data in these fields is expected to correspond to attributes and values.

Codable is a field with 'open ended content' like qualitative data. The data in these fields should typically be the subject of theme coding.

Datasets can only be created when data is imported. Data is arranged in the form of a matrix where rows are records and columns are fields. Typically, respondents are rows, columns are questions and cells are answers.

Importing Datasets

Structured data of any origin can be imported into NVivo so long as it meets the criteria described above:
1. Go to **External Data | Import | Dataset**.
   Default folder is **Internals**.
   Go to 5.

alternatively
1. Click [**Sources**] in Area 1.
2. Select the **Internals** folder in Area 2 or its subfolder.
3. Go to **External Data | Import | Dataset**.
   Go to 5.

alternatively
3. Click on any empty space in Area 3.
4. Right-click and select **Import Internals → Import Dataset...**

> **Tip:** An easy way to convert an Excel worksheet to text is:
> 1. Select the whole worksheet
> 2. Copy
> 3. Open Notepad
> 4. Paste into Notepad
> 5. Save with a new name

The **Import Dataset Wizard – Step 1** appears:

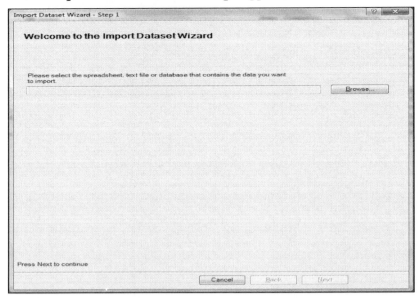

5   With the [**Browse**] button you will find the data file that you want to import. The file browser only displays file formats that can be imported as Datasets.
6   Confirm with [**Open**].
7   Click [**Next**].

The **Import Dataset Wizard** – **Step 2** appears:

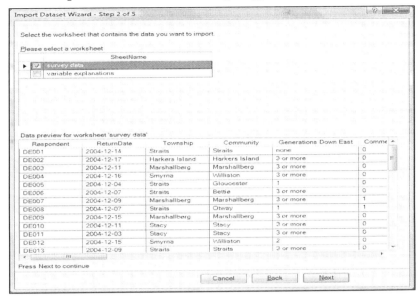

The upper section of the dialog box, SheetName, displays the two sheets of the Excel workbook: survey data and variable explanations. Select a sheet and its contents are displayed under Data Preview. The first 25 records of each are displayed. We select the sheet *survey data*.

8   Click [**Next**].

The **Import Dataset Wizard** – **Step 3** appears:

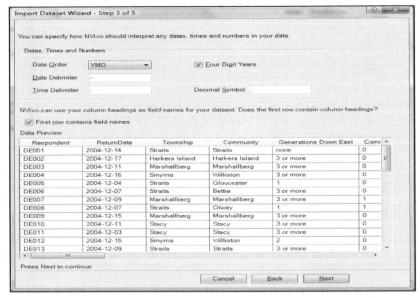

You can verify the Time and Date formats and the Decimal symbol against the information displayed in the Data Preview.

It is important that the field names of imported data are only in the first row. Certain datasheets have field names in two rows and if so then the two rows must be merged. If you uncheck the option *First row contains field names* the row instead will contain column numbers.

9   Click [**Next**].

The **Import Dataset Wizard – Step 4** appears:

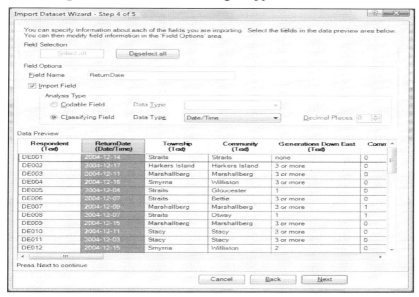

You must assign each column as either a *Codable Field* or as a *Classifying Field*. Select one column at a time by clicking the column head (or browse with **[Right]** or **[Left]** on the keyboard) in the Data Preview section. Use the options under *Analysis Type*. The default mode is Classifying for all columns. Unchecking the *Import Field option* for a certain column prevents its import.

10   Click **[Next]**.

The **Import Dataset Wizard** – **Step 5** appears:

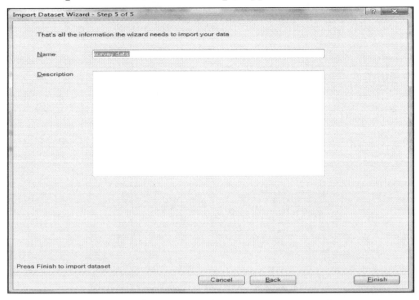

11   Type a name (compulsory) and a description (optional). Confirm with **[Finish]**.

A successful import creates a Dataset and when it opens in Area 4 and view mode *Table* it appears like this:

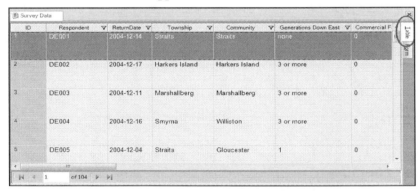

NVivo has created a new leftmost column, ID. A Dataset cannot be edited nor can you create or delete rows or columns. The buttons down left are for browse buttons between records.

View mode *Form* displays one record at a time:

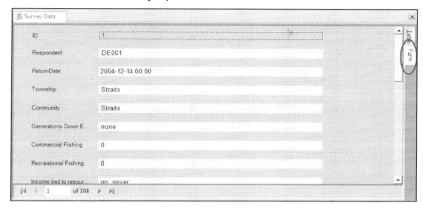

Classifying fields have a grey background and Codable fields a white background, like here in view mode *Table*:

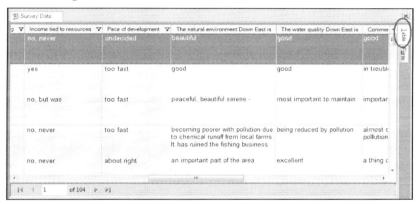

Or here in view mode *Form*:

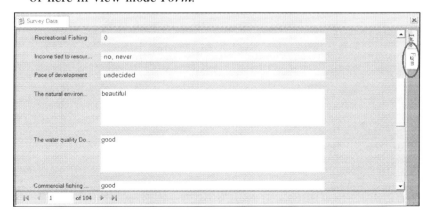

In the **Dataset Properties** dialog box you can do certain modifications to a Dataset's presentation:

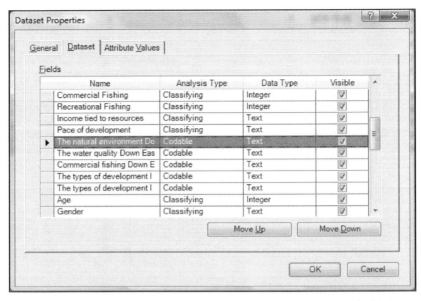

You can change names of a field, hide a field or move a field, but you cannot change Analysis Type or Data Type.

Alternatively, such modifications can also be made directly in a Dataset, view mode *Table*. All rules are as described for a Classification Sheet (Chapter 11, Classifications), and for Matrices (see page 201 and onwards), including the use of filters apply to Datasets.

Exporting Datasets

Datasets can be exported like other Project Items:
1. Click [**Sources**] in Area 1.
2. Select the **Internals** folder in Area 2 or its subfolder.
3. Select the Dataset in Area 3 that you want to export.
4. Go to **External Data | Export | Export → Export Dataset...** or right-click and select **Export → Export Dataset...** or [**Ctrl**] + [**Shift**] + [**E**].

The **Export Options** dialog box now appears.

5. Select applicable options and click [**OK**]. Then a file browser opens and you can decide file name, file location, and file type. Possible file formats are: Excel, .TXT and HTML.
6. Confirm with [**Save**].

Coding Datasets

Coding Datasets applies all the common rules: select text in codeable fields, right-click and select **Code Selection → Code Selection At New Node** or **Code Selection At Existing Nodes**.

All coding in a Dataset can be explored like in other Project Items including coding stripes and highlighting.

Autocoding Datasets

Autocoding Datasets is the opportunity to use Nodes to provide a structure to your Dataset content when you import it into NVivo. Autocoding Datasets is closely related to Autocoding Social Media Datasets (Chapter 18, see page 266).

1. Select a Dataset in Area 3 that you want to auto code or click in the open Dataset in Area 4.
2. Go to **Analyze | Coding | Auto Code**.

The **Auto Code Dataset Wizard – Step 1** appears:

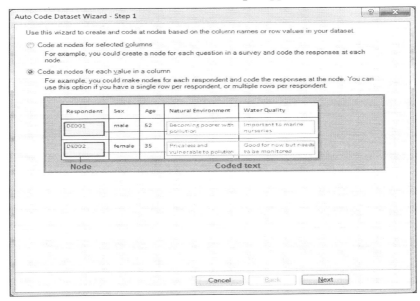

3. Select *Code at Nodes for each row,* then click [**Next**].

The **Auto Code Dataset Wizard – Step 2** appears:

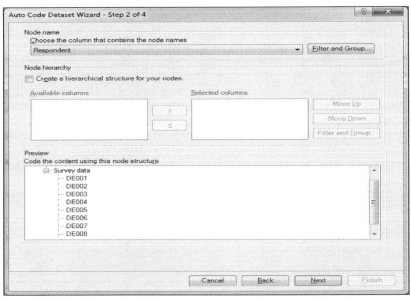

We select Respondent from the *Choose the column that contains the node names* drop-down list. No node hierarchy is needed.

4   Click **[Next]**.

The **Auto Code Dataset Wizard – Step 3** appears:

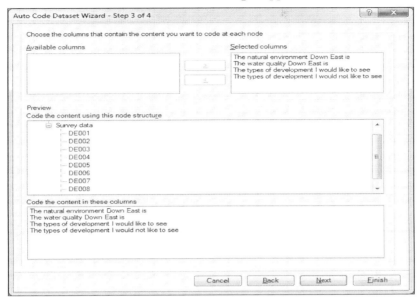

5 Select the fields that shall be coded from *Available Columns*, then click [>] and the fields are transferred to *Selected Columns*.
6 Click [**Next**].

The **Auto code Dataset Wizard – Step 4** appears:

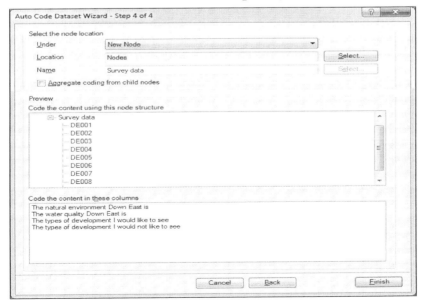

7   You can now decide the name and location of the new Nodes. In our example the location is **Nodes\\Survey Data**.
8   Confirm with [**Finish**].

The result is a collection of case nodes, one node for each respondent, showing in Area 3:

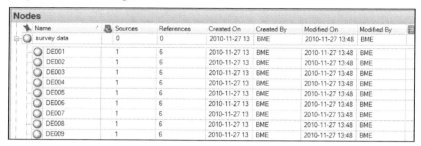

The next logical step would be to autocode our dataset in relation to the columns we have classified as Codable. Each such column will constitute a theme node. We return onece more to the Autocode command. This time we select *Code at nodes for selected columns* in the **Auto Code Dataset Wizard – Step 1:**.

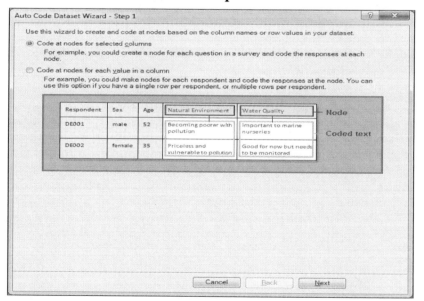

From here we proceed step by step similar to what we did when autocodning the rows.

Classifying Datasets

From a Dataset you can create and classify Nodes based on the Classifying fields. From the beginning there must be at least one existing Node Classification in the project.

If we classify existing Nodes (created from autocoding) these Nodes must be classified using the same Classification (possibly without Attributes and Values) applied in Stage 2 below.

1. Select all Nodes in Area 3 that you want to classify. Use **[Ctrl]** + **[A]** or select the first Node in the list, then hold down **[Shift]** and click the last Node in the list.
2. Right-click and select **Classification** → <**Classification Name**>.

The classification from the Dataset is then carried out like this:

1. Select the Dataset in Area 3 with the data that you want to use for classifying the Nodes in question
   or click on the open Dataset in Area 4.
2. Go to **Create | Classifications** → **Classify Nodes from Dataset**
   or right-click and select **Classify Nodes from Dataset**.

The **Classify Nodes from Dataset Wizard** – **Step 1** appears:

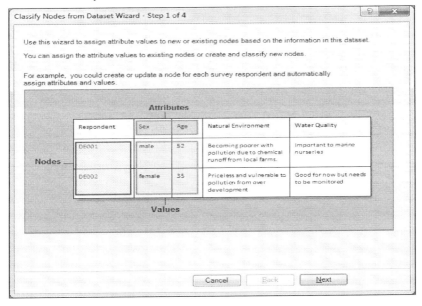

3. Click **[Next]**.

The **Classify Nodes from Dataset Wizard – Step 2** appears:

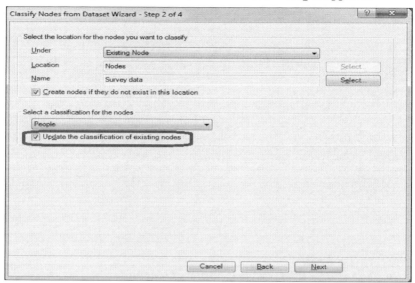

In our example we will classify the Nodes that were created with Autocoding. Therefore it is important to check *Update the classification of existing Nodes*.

4   Click [**Next**].

The **Classify Nodes from Dataset Wizard – Step 3** appears:

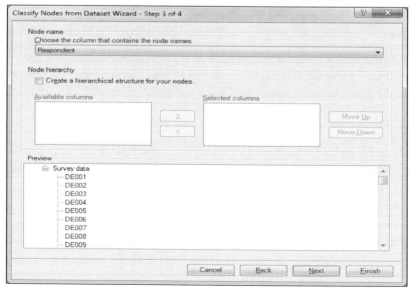

5   We select the column *Respondent* to create the Nodes. Click [**Next**].

254

The **Classify Nodes from Dataset Wizard – Step 4** appears:

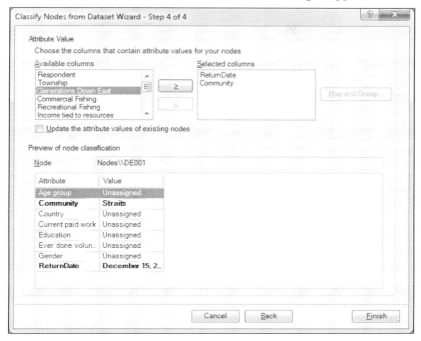

All Classifying fields are listed in the left box, *Available columns*. Use [>] to bring over the fields to the *Selected columns* box. In the section Preview the result from the topmost Node is displayed.

6   Click [**Finish**].

Mapping and Grouping

We return to the **Classify Nodes from Dataset Wizard – Step 4** above. The [**Map and Group**] button can be used to move (or map) the content from one column to another. There is also an option to group discrete numerical values as intervals, typically discrete ages of people to age groups:

1   In **Classify Nodes from Dataset Wizard – Step 4** the field *Age* has been moved to the right box, Selected columns.

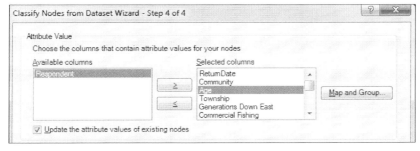

2   Highlight *Age* and click [**Map and Group**].

The **Mapping and Grouping Options** dialog box appears:

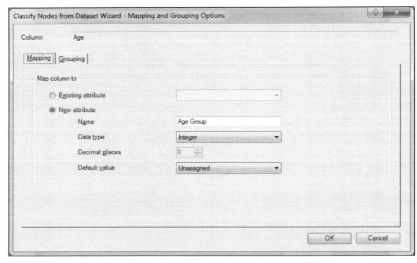

3   Select *New Attribute* which we call *Age Group*. Click on the **Grouping** tab.

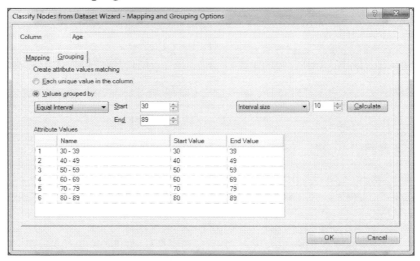

4   You can now decide the size of the interval. You can choose between *Equal Interval, Standard Deviation* or *User-defined Interval*. Confirm with **[OK]** and you will return to **Classify Nodes from Dataset Wizard – Step 4**.

18. INTERNET AND SOCIAL MEDIA

Arguably the most significant upgrade for NVivo 10 is the ability to import and handle data from internet web pages and social media sites like LinkedIn, Facebook and Twitter. NVivo 10 also features full integration with Evernote and OneNote, the popular cloud-based notetaking/archiving services that will also be discussed in the next chapters.

Introducing NCapture

NCapture is a browser plugin that is delivered and installed with NVivo 10. NCapture exports web content into files called *web data packages* (a .vcx file) that you will import into NVivo. NCapture allows you to export any website including the website's text, images and hyperlinks. Websites import into NVivo as PDF sources. NCapture also allows you to export data from LinkedIn, Facebook and Twitter. Social media data can also import into NVivo as a PDF source, but more importantly social media data can also be imported as an NVivo Dataset. Presently, NCapture is available as addins with Internet Explorer and Google Chrome, but other browsers may be available to work with NCapture as a part of future updates.

Exporting websites with NCapture

Like most software commands, capturing web data with NCapture in a web-browser can happen in three ways:

1. Select the **NCapture icon** on the toolbar of the web-browser:

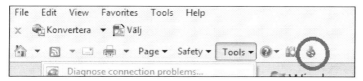

2. Select **NCapture for NVivo** from the Tools menu:

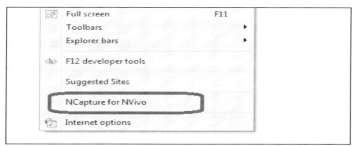

3. Right-click anywhere in your web-browser and select **NCapture for NVivo**.

Importantly, you can export numerous web data packages during your online research. NVivo does not require you to import your web data until you're ready. When you select to import a website to NVivo, the NCapture dialog box appears in the web-browser:

For websites, your Source Type will be Web Page as PDF by default as NVivo can currently only create website data packages that will be imported as PDF sources. But a variety of useful options are available for you to customize how your web data package can be imported into NVivo:

Source name will be the name of your new PDF source –the website's name will be the default here.

The *Description/Memo* tab allows you to type custom text that you want to add into *description field* of the PDF Source Item or a newly created *linked memo* with the same name as the Source Item. Which of these options works best for you will depend on your project – remember, linked memo content can be searched and coded; description field text cannot. Code at Nodes: You can type the names of any number of new or existing Nodes here. NCapture only has the ability to code web content at Nodes located in the Nodes folder. The imported PDF Source Item will be 100% coded at Nodes entered in the *Code at Nodes* field.

Importing Websites with NCapture

After NCapture exports your data, you will need to retrieve and import the newly created web data package (.vcx) file(s). When you have returned to NVivo:

1. Go to **External Data | Import | From Other Sources → From NCapture...**
   Default folder is **Internals**.
   Go to 5.

alternatively

1. Click on [**Sources**] in Area 1
2. Select the **Internals** folder in Area 2 or its subfolder
3. Go to **External Data | Import | From Other Sources → From NCapture...**
   Go to 5.

alternatively

3. Click on any empty space in Area 3.
4. Right-click and select **Import → Import from NCapture...**
   The **Import From NCapture** dialog box appears:

At the bottom of this dialog box, all recently imported items from NCapture are listed. NVivo will detect if there are web data packages that you have already imported, and so the default selection is All captures not previously imported. You can also select to import All captures or Selected captures.

5. Click [**Import**] and the result will be as follows:

The sample PDF Source item below is an export from the website 'Conflicts in Africa – Global Issues'. You'll notice that the webpage title is the same as the name of the PDF source file. Imported NCapture websites are classified with the Source Classification 'Reference'. Values are inserted by default for the following Attributes: Reference Type, Title, keywords, URL and Access Date. As you'll recall from our sample export image above, this entire PDF source will be coded at two Nodes, *Africa* and *Conflicts* and a linked memo will have been created sharing the PDF source's name, *Conflicts in Africa – Global Issues.*

Now you can open the source and hyperlinks are clickable like in any PDF item by using **[Ctrl]** + click

Social Media Data and NCapture

NCapture can also be used to capture a wealth of data from Facebook, Twitter, and LinkedIn. Social media web data packages can be created as PDF sources or Datasets, which will be the focus of our description below.

Due to each social media site's unique structure, NCapture captures different types of data from each site. While a summary of the complete functionality of Facebook, Twitter, and LinkedIn is beyond our purposes here, we will provide an explanation of the types of data you can capture from each site. Importantly, your ability to capture social media data is contingent on the privacy settings of the individual or group whose data you are interested in capturing (e.g., some Twitter users may require you to be their Follower before you can capture their Twitter data).

Importantly, you can use NCapture to gather social media data over a period of time and then easily update the data later. When you import web data packages containing social media data, by default, new data will be merged with old data so long as the original social media properties (e.g., hashtags, usernames, etc.) remains the same.

NCapture for Facebook Data

NCapture allows you to capture Facebook wall posts and data about their authors. Whether from an individual's Facebook wall (e.g., Allan McDougall), a Group wall (e.g., the Stockholm Sailing Club), or a Page wall (e.g., QSR International), NCapture can export wall posts, tags, photos, hyperlinks, link captions, link descriptions, number of 'likes', comments, comment 'likes', dates and times of posts and comments. Further, NCapture can export authors' names, genders, birthdays, locations, relationship statuses, bios, religions, and hometowns.

NCapture for Twitter Data

NCapture allows you to capture Twitter tweets and data about their authors. Unlike Facebook, which is largely based on users being connected as 'friends' or as fans who 'like' a specific page, Twitter profiles and their attendant tweets are (typically) publically available. As a result, along with individual user streams, full Twitter searches can also be exported with NCapture. Whether for user streams or search results, NCapture can export tweets along with their attendant usernames, mentions (usernames within tweets), hashtags (user-driven keywords), timestamps, locations, hyperlinks (if any), retweets (reposts by other users), and usernames of any 'retweeters'. Unlike NCapture's ability to export demographic data from Facebook, NCapture for Twitter captures data associated with a user's influence level (or klout), such as number of tweets, number of followers, and the number of users they are following.

NCapture for LinkedIn Data

Capturing social media data from LinkedIn is more similar to Facebook than Twitter. NCapture allows you to capture discussions and comments from LinkedIn groups, rather than individual user's LinkedIn profile pages. From any given LinkedIn group, NCapture exports information about each post and author. LinkedIn Datasets contain each post's title, attachments, timestamp, 'likes', and headline, as well as the authors name, location, industry, date of birth, and number of connections.

Tip: Although you can't export LinkedIn users' profile data as a dataset with NCapture, you can still export user profiles as a PDF source. While unstructured, these PDF source can still be searched and coded after you import them into NVivo.

Exporting social media data with NCapture

Once you have found social media data that you need then activate NCapture from Internet Explorer or Google Chrome as described above. For social media web data packages, the default source type is a Dataset. You can change from Dataset to PDF with the drop-down list.

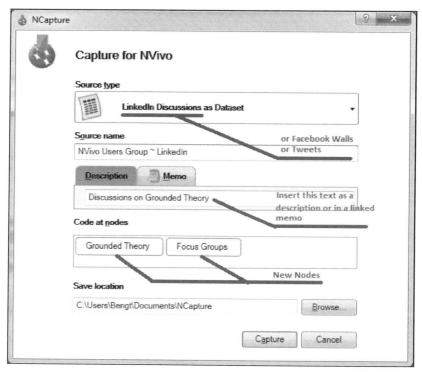

Like exporting websites with NCapture, when you export social media data you can create an item description, linked memo, and Nodes. When you have filled in the NCapture dialog box, then click **[Capture]**.

Importing Social Media Data with NCapture

Now that you've exported your social media data to a web data package, it's time to import:
    1   Go to **External Data | Import | From Other Sources → From NCapture...**
        Default folder is **Internals**.
        Go to 5.
alternatively
    1   Click on **[Sources]** in Area 1.

2   Select the **Internals** folder in Area 2 or its subfolder
3   Go to **External Data | Import | From Other Sources → From NCapture...**
    Go to 5.

alternatively

3   Click on any empty space in Area 3.
4   Right-click and select **Import → Import from NCapture...**
    The **Import From NCapture** dialog box appears:

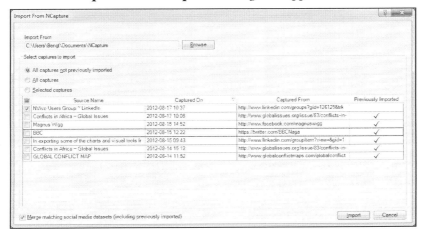

5   Like website data, at the bottom of this dialog box all recently imported items from NCapture are listed. NVivo will detect if there are web data packages that you have already imported, and so the default selection is *All captures not previously imported*.
6   Click **[Import]** and the result will be as follows:

The sample Dataset Source Item below is an export from the 'NVivo Users Group on LinkedIn'. You'll notice the Dataset Source Item contains the LinkedIn group name. Imported NCapture social media data is classified with the Source Classification 'Reference'. Values are inserted by default for the following Attributes: Reference Type, Title, keywords, URL and Access Date. As you'll recall from our sample export image above, this entire Dataset source will be coded at two Nodes, *Grounded Theory* and *Focus Groups*, and a linked memo has been created sharing the Dataset source's name, *NVivo Users Group on LinkedIn*.

Now you can open the Source and view it in several modes: Table (default), Form, or as a Cluster

When NCapture exports a photo from Facebook, the photos are stored as separate picture source file in the same folder as the Facebook dataset when imported. The icon shown is a *source shortcut* and a placeholder that allows you to easily navigate between your dataset and the photos (or other non-database files) you've imported.

Analysis. The latter mode is unique for Datasets from Social Media and usually clusters Usernames.

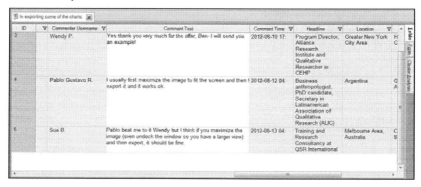

Working with Social Media Datasets

What makes working with social media Dataset sources so exciting, like working with any Dataset source, is your ability to easily edit, customize, and survey your structured data through Dataset Properties:

1. Click on **[Sources]** in Area 1.
2. Select the **Internals** folder in Area 2 or its subfolder.
3. Select the dataset that you want to edit.
4. Go to **Home | Properties | Dataset Properties**
   or right-click and select **Dataset Properties**
   or **[Ctrl} + [Shift] + [P]**.

alternatively

4. From any open any dataset, go to **Home | Properties | Dataset Properties**.

Within the **General** tab you can change the name and description of your social media Dataset.

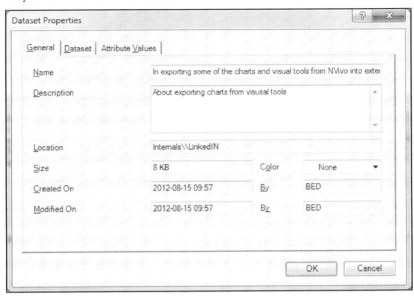

The **Dataset** tab allows you to view your Dataset fields and move some fields up or down. Upon importing your social media web data package, NVivo has already decided which columns are Classifying (a limited set of options) and which columns are Codable (editable text) respectively. Within this tab you can uncheck Visible on any row with data you deem unnecessary, like some demographics (e.g., Hometown for Facebook data).

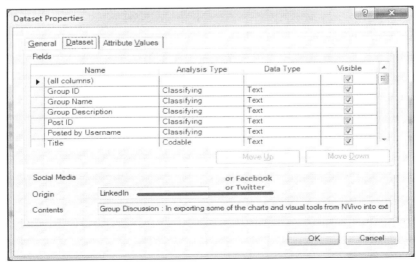

The **Attribute Values** tab allows you to view default Attribute Value information. You can select a custom Classification if that is useful for your project:

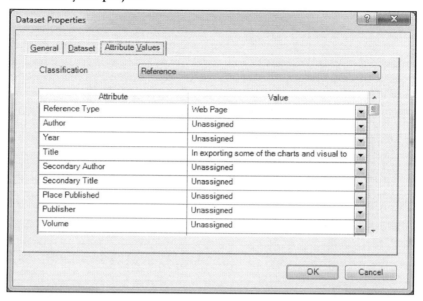

Analyzing Social Media Datasets

A number of exciting methods for analyzing social media Datasets exist in NVivo 10. Like any open Dataset source, you can search for patterns in your data by hiding, sorting, or filtering rows and columns. More advanced analysis functions such as Word Frequency Queries and Text Search Queries can offer insight into some themes in your data as well. Further, visualizations of social media data can be achieved using chart functions (see page 169).

Autocoding a Dataset from Social Media

Perhaps the most useful tool for social media web data is autocoding.
1. Select a Dataset in Area 3 that you want to auto code.
2. Go to **Analyze | Coding | Auto Code**
   or right-click and select **Auto Code...**

The **Auto Code Dataset Wizard – Step 1** appears:

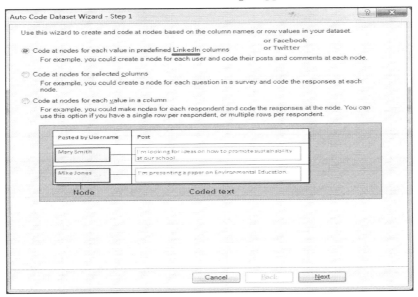

The first option, *Code at Nodes for each value in predefined LinkedIn (or Facebook or Twitter) columns* is unique for importing social media compared to other types of Datasets. For example, using the Auto Code Dataset Wizard we can create Nodes containing all of the content generated by one user, or all of the comments generated during a group discussion.

3 Click [**Next**].

The **Auto Code Dataset Wizard** - **Step 2** appears:

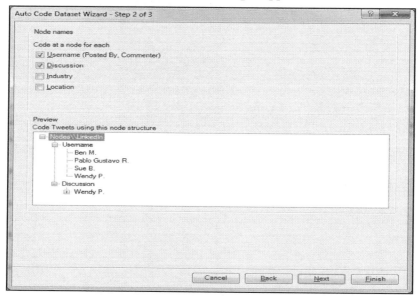

The **Auto Code Dataset Wizard** provides a preview of the resulting Node structure that your audo coding will generate. As you can see in the above image, by selecting to code data at Username and Discussion, Node hierarchies will be created where all of the data generated by each user will be coded into a Node named after that user.

4   Click [**Next**].

The **Auto Code Dataset Wizard** - **Step 3** appears:

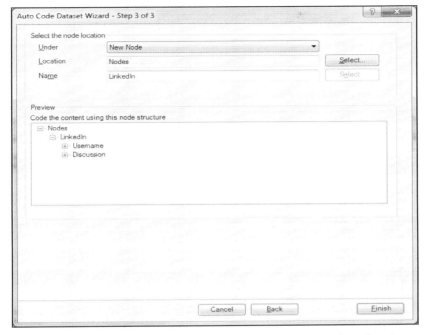

Default settings will create Nodes and apply a Node Classification named LinkedIn User, Facebook User or Twitter User which classifies the users under respective parent Node LinkedIn/Username, Facebook/Username or Twitter/Username. Also other Nodes have been created under the parent Nodes LinkedIn (Discussion), Facebook (Conversation) or Twitter (Hashtags).

Like other NVivo Datasets, the auto code wizard can also allow you to create Nodes based on the columns (e.g., all hashtags would become Nodes within a parent Node called hashtags) and cell values of the Dataset (e.g., all unique hashtags become their own Nodes with each occurrence auto coded).

Privacy levels can vary using social media so contact QSR Support if you have any problems with importing social media data.

5   Finally click [**Finish**].

The result from these operations is not only a easy-to-handle dataset, a set of casenodes (Usernames) and a set of theme nodes (Comment text, Post, Title) but also a Source Classification and a Node Classification.

The Source Classificationen **Reference** was created when data from NCapture was imported as a dataset:

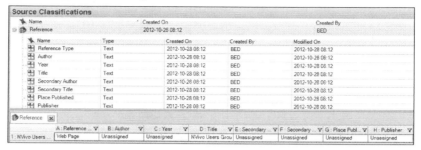

The Node Classification **LinkedIn User** was created when our dataset was autocoded with reference to Username:

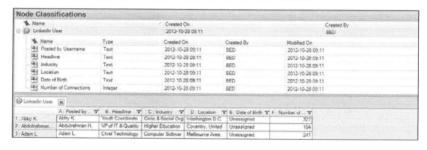

Installing NCapture

For Internet Explorer:
1   Download NCapture.IE.exe from QSR's web page.
2   Close Internet Explorer.
3   Launch NCapture.IE.exe and follow the prompts to complete the installation.

For Google Chrome:
1   Run Google Chrome.
2   Find the link with the installation guide on QSR's web page.
3   Follow the prompts to complete the installtion.

Check your Version of NCapture

For Internet Explorer:
Go to **Tools → Manage Add-ons**
View the version number for **NCapture for NVivo** in the list.

For Google Chrome:
Go to **Tools → Extensions**
View the version number for **NCapture for NVivo** in the list.

19. USING EVERNOTE WITH NVIVO

One of the new features in NVivo 10 that we are most excited about is the software's native capacity for importing data from Evernote. If you aren't already an Evernote user, it is a software suite designed for note taking and archiving. Evernote is cloud-based, which means that your notes are stored on an online server rather than on a local hard drive on a computer. The name Evernote implies that your notes will be archived 'forever' on the Evernote server. Do you have privacy concerns about uploading data to a cloud-based service like Evernote? Check with your institution's IT group or ethics office to find out what kind of data you are allowed to capture on Evernote and other popular cloud-based resources like Dropbox, Skydrive, and Google Drive.

Evernote for Data Collection

Evernote has gained popularity because of its functionality for smart phone users. As a note taking program that is available on Blackberry, iPhone, iPad, and Android smart phones.

For qualitative researchers, a smart phone with Evernote installed offers a range of new possibilities for data collection. From the same device, researchers can record audio and video, take photos, capture web data, and easily upload all of these multimedia resources into NVivo 10.

Exporting Notes from Evernote

Evernote notes must be exported as an .ENEX file from the Evernote client, which is basically a XML format, for further import to NVivo. While a tutorial on how to use Evernote is beyond our purposes here, we have included a screen shot below displaying the File menu in Evernote. In order to export this picture note, simply go to **File → Export → Export as a file in ENEX format**:

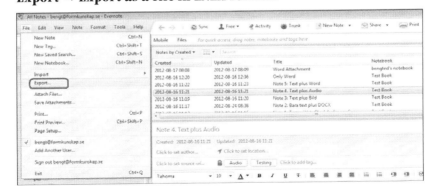

Importing Evernote Notes into NVivo

Remember you can batch import a set of notes or an entire Evernote notebook to an .ENEX file. Once you save your .ENEX file you can easily import its content into NVivo:
1   Go to **External Data | Import | From Other Sources → From Evernote...**
    Default folder is **Internals**.
    Go to 5.

alternatively
1   Click on **[Sources]** in Area 1.
2   Select the **Internals** folder in Area 2 or its subfolder
3   Go to **External Data | Import | From Other Sources → From Evernote...**
    Go to 5.

alternatively
3   Click on any empty space in Area 3.
4   Right-click and select **Import → Import from Evernote...**
The **Import from Evernote** dialog box appears:

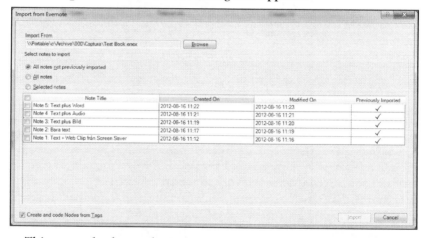

This example shows the import dialog when a notebook has been exported. The bottom of the dialog box contains a list of each individual note included in the .ENEX file. NVivo will detect if there are notes in the .ENEX that you have already imported, and so the default selection is to import *All notes not previously imported*. Alternatively, you can select import *All notes* or import *Selected notes*. Select the notes that you want to import.

5   Click **[Import]** and the result will import.

Evernote Note Formats in NVivo

Not all Evernote data will be imported as internal sources, so it is a good idea to familiarize yourself with the following list of Evernote note types and their attendant locations in NVivo:

- Evernote text notes will become a document Source in the internal Sources folder.
- Evernote notes with file attachments (e.g., PDFs, photos, images, audio files, or video files) will retain their file types. Any text that accompanies these notes will become a linked memo associated with the newly imported Source Item.
- Evernote web clippings (i.e., web page which has been saved to Evernote via an Evernote Web Clipper) will be imported as PDF Source Item.

Autocoding your Evernote Tags

Some Evernote users using note tagging as a way of linking notes together with broad categories for later reference and searching. Evernote tags in this sense are similar to NVivo Nodes. A nice feature for importing your tagged Evernote notes allows you to convert your tags to Nodes when you import the Evernote note. These Nodes are created in the Nodes folder (if they do not already exist). The Node will code the entire imported source. If you do not want to create Nodes when your notes are imported, clear the *Create and code Nodes from Tags* check box uin the dialog box **Import from Evernote**.

20. USING ONENOTE WITH NVIVO

One of the new features in NVivo 10 that we are most excited about is the software's native capacity for importing data from OneNote. If you aren't already an OneNote user, it is a software suite designed for note taking and archiving. OneNote is cloud-based, which means that your notes are stored on an online server rather than on a local hard drive on a computer.

Exporting Notes from OneNote

Export to NVivo is made by an Addin component normally installed when NVivo 10 is installed. OneNote pages must be exported as a .NVOZ file from the OneNote client, which is basically a XML format, for further import to NVivo. While a tutorial on how to use OneNote is beyond our purposes here, we have included a screen shot below displaying the **Share** ribbon in OneNote. In order to export these pages, simply go to **Share | NVivo | Export**:

Then the **Export for NVivo** dialog box appears.

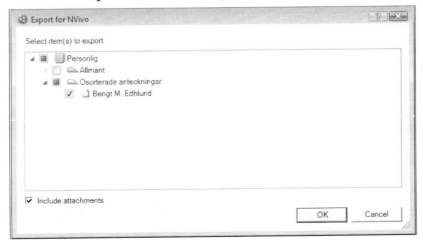

Importing OneNote Notes into NVivo

Once you save your .NVOZ file you can easily import its content into NVivo:

1. Go to **External Data | Import | From Other Sources → From OneNote...**
   Default folder is **Internals**.
   Go to 5.

alternatively

1. Click on **[Sources]** in Area 1.
2. Select the **Internals** folder in Area 2 or its subfolder
3. Go to **External Data | Import | From Other Sources → From OneNote...**
   Go to 5.

alternatively

3. Click on any empty space in Area 3.
4. Right-click and select **Import → Import from OneNote...**

The **Import from OneNote** dialog box appears:

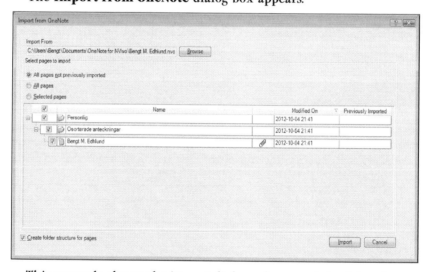

This example shows the import dialog when a notebook has been exported. The bottom of the dialog box contains a list of each individual note included in the .NVOZ file. NVivo will detect if there are notes in the .NVOZ that you have already imported, and so the default selection is to import *All pages not previously imported*. Alternatively, you can select import *All pages* or import *Selected pages*. Select the pages that you want to import.

5. Click **[Import]** and the result will import.

OneNote Note Formats in NVivo

Not all OneNote data will be imported as internal sources, so it is a good idea to familiarize yourself with the following list of OneNote note types and their attendant locations in NVivo:

- OneNote text notes will become a document Source in the internal Sources folder.
- OneNote notes with file attachments (e.g., PDFs, photos, images, audio files, or video files) will retain their file types.

Installing NVivo Addin for OneNote

1. Download NVivoAddIn.OneNote.exe from QSR's website.
2. Make sure that OneNote is not running.
3. Click **Run**.

Check whether NVivo Addin for OneNote is Installed

For OneNote 2010:
1. Go to **File → Options → Add-Ins**
2. Check if Export for NVivo is on the list.

For OneNote 2007:
1. Go to **Tools → Options → Add-Ins**
2. Check if Export for NVivo is on the list.

21. FINDING AND SORTING PROJECT ITEMS

This chapter is about how to find Project Items. The finding tools in NVivo are *Find* and *Advanced Find*. Another useful function for finding relations between items is *Group Queries*, which is dealt with on page 206. The results of these functions are lists of shortcuts to the found items.

Find

The bar **Find** is always just above the List View heading for Area 3. This bar can be hidden or unhidden with **View | Workspace | Find** which is a toggling function. The easy function **Find Now** is used for finding names of documents, memos or Nodes, not their contents.

1. At **Look for** you type a whole word or a fragment of a word that is part of the name of an item. Here is free text search applied (not whole words, not case sensitive).
2. The drop-down list **Search In** is used to select to which folder or folders the search shall be restricted.
3. Click [**Find Now**].

The result is a list of shortcuts in Area 3. A shortcut is indicated by a small arrow in the bottom-right corner of the icon. The list cannot be saved but you can create a **Set** from selected items from the list (see page 29).

Advanced Find

Advanced Find gives increased specificity to any given search.
1   In the **Find** bar go to **Advanced Find**
    or key command **[Ctrl]** + **[Shift]** + **[F]**.
The **Advanced Find** dialog box are shown.
The drop-down list **Look For** has the following options:
- Sources
- Documents
- Audios
- Videos
- Pictures
- Datasets
- PDFs
- Externals
- Memos
- Framework Matrices
- Nodes
- Relationships
- Node Matrices
- Source Classifications
- Node Classifications
- Attributes
- Relationship Types
- Sets
- Queries
- Results
- Reports
- Extracts
- Models
- All

As an example of Advanced Find options, you can limit a text search to just the Description box of a certain type of Project Item.

The Intermediate Tab

The **Intermediate** and **Advanced** tabs are independent of each other. Below is the **Intermediate** tab of the **Advanced Find** dialog box:

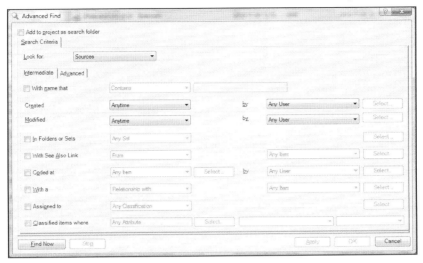

As soon as any option in the Intermediate tab has been chosen the corresponding **[Select...]** button is activated and opens the **Select Project Items** dialog box. The exact shape of this dialog box is determined by the selected option.

This function can be used to create a list with items matching certain criteria, like:
- Nodes created *last week*
- Nodes that are *Male*
- Memos with a 'See Also Link' from the Node *Adventure*
- Documents that are coded at the Node *Passionate*
- Nodes that code the document *Volunteers Group 1*
- Sets containing *Nodes*

The Advanced Tab

The **Advanced** tab offers other types of criteria:

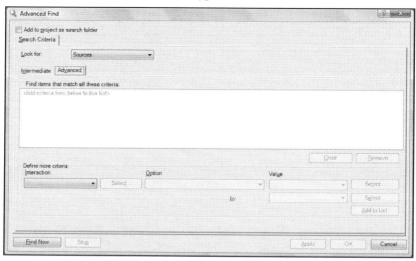

The **Interaction** drop-down list depends on the type of item that you have selected at **Look for**. For example, if *Documents* is selected the drop-down list has the following options:

- Document
- Name
- Description
- Created
- Modified
- Size (MB)
- Attribute

1   Select *Nodes* from the **Look for:** drop-down list. In the section *Define more criteria* the drop-down list now has options specifically for Nodes.
     In this case, select:
     Age Group / equals value / 50-59 and the dialog box looks like this:

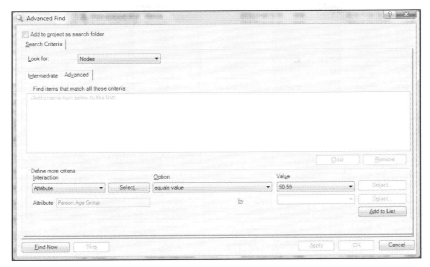

2   Click [**Add to List**] and the search criteria moves to the box **Find items that match all these criteria**.

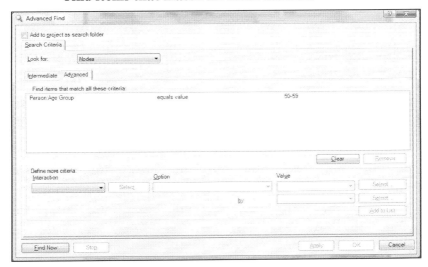

3   You can now add another criterion for example a limitation to women. Then again click [**Add to List**].

4   Finally, the search is done with [**Find Now**] and the result looks like this:

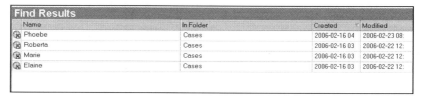

The result is a list of shortcuts that match the search criteria. This list can be stored in a subfolder of the **Search Folder**. This subfolder can be created by checking *Add to project as search folder* in the **Advanced Find** dialog box. The **New Search Folder** dialog box is shown. Type a name (compulsory) and a description (optional):

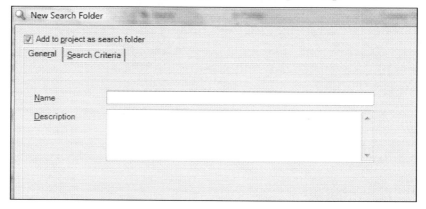

Click [**Folders**] in Area 1 and then you can open the folder **Search Folders** in Area 2 and then you can find the new folder. Click the folder and the whole list of shortcuts will appear in Area 3.

You can also create a set of selected items from this list (see page 29).

Sorting Items

This section applies to all items that can be viewed in a list usually in Area 3, but sometimes also in Area 4. For example, when a Node is opened in view mode Summary, a list is shown in Area 4.

1. Display a list of items in Area 3.
2. Go to **Layout | Sort & Filter | Sort By** → <select>.

The options offered depend on of the type of items in the list. Nodes, for example, can be arranged hierarchically, so for Nodes there is a special sorting option, Custom.

1. Display a list with Nodes in Area 3.
2. Go to **Layout | Sort & Filter | Sort By** → **Custom**.
3. Select the Node or Nodes that you want to move. If you want to move more than one Node they must be adjacent.
4. Go to **Layout | Rows & Columns | Row** → **Move Up/ Move Down**
   or **[Ctrl] + [Shift] + [U]/[Ctrl] + [Shift] + [D]**.

This sorting is automatically saved even if you temporarily change the sorting. You can always return to your Custom sorting:

1. Display a list of items in Area 3.
2. Go to **Layout | Sort & Filter | Sort By** → **Custom**.

This command is a toggling function. When you use the command again it sorts in the opposite order.

You can also use the column heads for sorting. Sorting by commands or sorting with column heads always adds a small triangle to the column head in question. Clicking again on this column head turns the sorting in the opposite order.

22. COLLABORATING WITH NVIVO 10

As technology and interdisciplinarity facilitate more and more complex qualitative studies, teamwork structures and procedures become increasingly important. NVivo allows several users to use the same project file provided that the file is opened by one user at a time. Alternatively, each member can work with his/her own project file that can be merged into a master file at a certain predefined occasion. The focus of this chapter focuses on how a team can operate using a single project file. The first half of this chapter explains some collaboration tools features in NVivo. The second half explains some general insights on collaborating with NVivo.

Collaborating on the same NVivo project can be arranged in a number of ways:
- Team members can use the same data but each individual creates his/her Nodes and codes accordingly – perhaps importing to a master project later.
- Team members use different data but use a common Node structure.
- Team members use both the same data and a common Node structure.

In cases where individual team members plan to merge their analytic progress into a master project file, merging projects is described on page 60. Review the options of the **Import Project** dialog box to find out how it can suite your needs. If Nodes with same names need to merge you can select Merge into existing item. Remember that Nodes and other items must have the same name and must be located on the same level of the folder structure before they can be merged successfully. Further, the contents of the Source Items must be identical.

NVivo includes several useful tools for collaborative data analysis:
- *View Coding Stripes by Selected Users* (or *View Substripes*).
- *View Coding by Users* in an open Node.
- *Coding Comparison Queries* for comparing two coders working with the same sources and Nodes. This is an important option that improves a project's validity and quantifies inter-rate reliability.

Current User

An important concept for teamwork in NVivo is the **Current User**. In **File → Options** and the **Application Options** dialog box, the **General** tab identifies that the current user. When a project is open you can change the current user. However, it is not possible to leave the Name and Initial boxes empty.

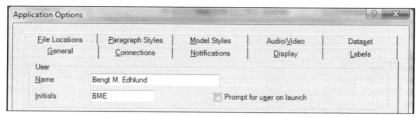

If you select the option Prompt for user on launch then the **Welcome to NVivo** dialog box is prompted each time NVivo is started:

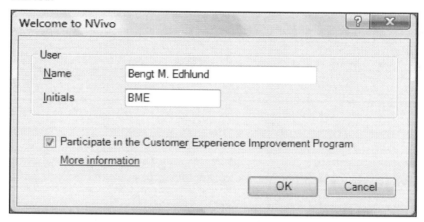

All users who have worked on the project are listed in **File → Info → Project Properties...** and in the **Project Properties** dialog box, under the **Users** tab:

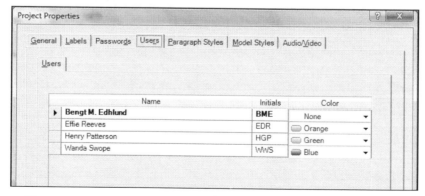

The current user is written in bold. The small triangle in the left column indicates the user who created the current project. In this box you cannot change the names but the initials. To the left in the status bar the current user is shown:

Initials are used to identify all Project Items created or modified by a certain user.

Viewing Coding by Users

NVivo allows you to to view the coding made by a certain member of a team:
1 Open the Node you wish to review.
2 Go to **View | Detail View | Node → Coding by Users → <select>**.
3 Choose any of the options *All Users, Current User, Selected User..., Select Users...*

The default setting is *All Users* and during a work session the selected option will remain. Selecting *Select Users* will show in **bold** the users of this Node. When a certain user has been selected a filter funnel symbol is shown in the status bar.

Viewing Coding Stripes

Coding stripes and sub-stripes can be used to display the coding that individual team-members have made (see page 167):
1 Open the Source Item you wish ro review.
2 Go to **View | Coding | Coding Stripes → Selected Items**.

The **Select Project Items** dialog box is shown. Only Nodes used to code the current item have names in **Bold**. When you select **Users** (and select individual users) one coding stripe per user is shown. When you point at one such stripe the names of the Nodes at which each user has coded will show. The substripes are Nodes:

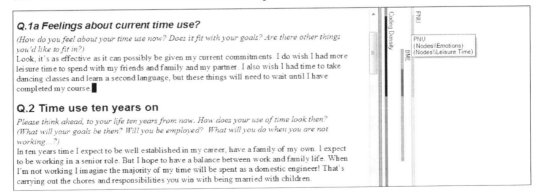

289

When you instead select **Nodes** (and select some individual nodes) one coding stripe per Node is shown. When you point at one such stripe the names of the users who has coded at those Nodes will show. The sub-stripes are Users.

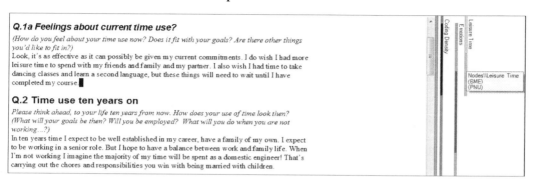

You can also display sub-stripes at the same time as the normal coding stipes by pointing at a coding stripe, right-click and select **Show substripes → More sub-stripes...** and select one or more sub-stripes that you want to show. Here are some sub-stripes for Users:

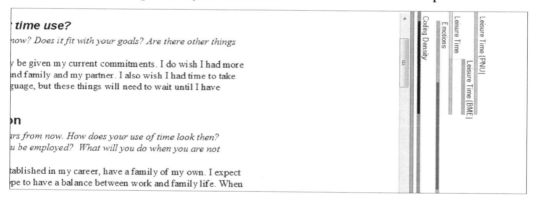

Coding Comparison Query

For projects interested in studying inter-rater reliability it is possible to compare how two people or two groups of people have coded the same material. This is possible provided that the same source material and the same Node structure have been used:

1. Go to **Query | Create | New Query → Coding Comparison...**
   Default folder is **Queries**.
   Go to 5.

alternatively
1   Click **[Queries]** in Area 1.
2   Select the **Queries** folder in Area 2 or its subfolder.
3   Go to **Query | Create | New Query → Coding Comparison...**
     Go to 5.
alternatively
3   Click on any empty space in Area 3.
4   Right-click and select **New Query → Coding Comparison...**
The **Coding Comparison Query** dialog box appears:

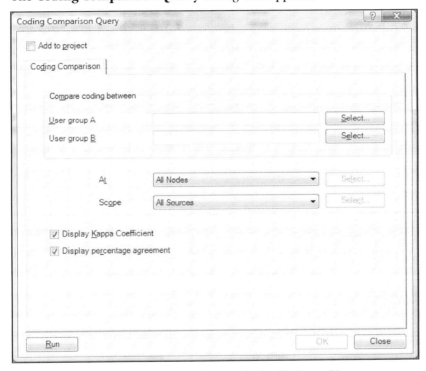

5   Define User group A and B with the **[Select...]** buttons which give access to all users that have been working in the project.
6   The **At** drop-down list determines the Node or Nodes that will be compared.
7   The **Scope** drop-down list determines the Source Item or items that will be compared.
8   Select at least one of the options *Display Kappa Coefficient* or *Display percentage agreement*.
9   You can save the query by checking *Add To Project*.
10   Run the query with **[Run]**.

The result can look like this:

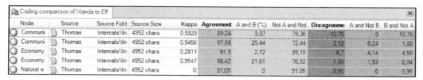

The percentage agreement columns indicate the following values:
- **Agreement Column** = sum of columns **A and B** and **Not A and Not B**.
- **A and B** = the percentage of data item content coded to the selected Node by both Project User Group A and Project User Group B.
- **Not A and Not B** = the percentage of data item content coded by neither Project User Group A and Project User Group B.
- Disagreement Column = sums of columns A and Not B and B and Not A.
- **A and Not B** = the percentage of data item content coded by Project User Group A and not coded by Project User Group B.
- **B and Not A** = the percentage of data item content coded by Project User Group B and not coded by Project User Group A.

From each row of the result from a Coding Comparison Query any *Node* can be analyzed like this:
1. Select a row from the list of results.
2. Go to **Home | Open | Open Node...**
   or right-click and select **Open Node...**
   or **[Ctrl] + [Shift] + [O]**.

Any Node that is opened from such list is showing the coding stripes and sub-stripes that belong to the users who are compared.

From each row of the result from a Coding Comparison Query any *Source Item* can be analyzed like this:
1. Select a row from the list of results.
2. Go to **Home | Open | Open Source...**
   or right-click and select **Open Source...**
   or double-click on the row.

Any Source Item that is opened from such a list shows the coding stripes and sub-stripes that belong to the users who are compared:

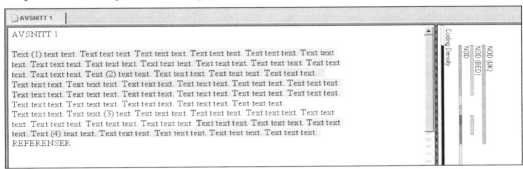

Coding stripes can always display the coding made by an individual user. This is made by pointing at a certain coding stripe, right-clicking and selecting **Show Sub-Stripes** and then selecting one or several users. Hiding sub-stripes is made using **Hide Sub-Stripes**.

Models and Reports

During team project meetings, modelling can be illustrative and useful. Any Node structure can easily be made understandable using NVivo Models (see the next chapter).

Reports are created by going to Explore | Reports | New Report. These reports can be used to study Nodes and coding made by the different team-members, see Chapter 25, Reports and Extracts.

Tips for Teamwork

Based on our years of experience working with hundreds of qualitative researchers using NVivo, we offer our colleagues the following tips for collaborating with NVivo:

- Appoint **an NVivo coordinator** for the research project.
- Set up **file name protocols**, read-only, storage locations, backup locations, file distribution and archiving.
- Set up **rules for audio and video files** like file formats and file distribution. For example, should you use embedded items or external files?
- Set up a **Node strategy**. Such a strategy can be communicated in a number of ways. It is easy to make a Node template in the form of a project without Source Items. Each Node should have 'instructions' written in the Node's Description field (max 512 characters) or in the form of a linked Memo, which is easier to write, read, print and code. The Node template can be distributed to team-members, saved with a new name and developed into a project in its own right. Importantly, the Node template's structure must not be modified by users. When new ideas are evolved, users should instead create new Nodes in addition to the Node template and create Memo Links.
- Determine how **Node Classifications** and **Source Classifications** will be applied. Such Nodes can be interviewees or other research items like places, professions, products, organizations, phenomena. In some situations it is useful to work with different classifications.
- Set up **rules for the master project** including protocols for merging and updating. Define a new project with a new name that clearly indicates that it is a merged project. Possibly a new set of user names will be defined for this purpose. Import one partial project at a time with **Import Project** and the option 'Merge into existing item'. Items with same name and same location will be merged.
- Hold periodic **team meetings** for the project. Such meetings should compare and analyze data (as described in this chapter), summarize discussions, and make decisions. Distribute minutes from each meeting.
- Assuming that the work has come to a stage where different members have submitted contributions to the project, make sure that the team has the standardized **usernames** when they work with their respective parts.

Continuing to work on a Merged Project
After exploring a merged project you have two options to proceed:
- Each user continues with the original individual projects and at a certain point of time you make a complete new merger – perhaps archiving the original merger.
- Each user continues to work on the merged project and archives the original individual portions.

We recommend continuing with the first option up to a certain point and then, if the team agrees, deciding to focus on the merged project later.

A Note on Cloud-computing
Some researchers we have worked with use cloud-based file sharing services like DropBox, SkyDrive and Google Drive as a working solution for collaborating on an NVivo Project. These services allow changes to the NVivo project file to be made across several computers using the 'cloud'. We recommend you turn off the live syncing features of these programs while you are running NVivo. We have been contacted by a number of colleagues and clients who have lost data while simultaneously using NVivo and syncing its attendant (.nvp) file. Again, cloud-based utilities can be useful for team collaboration, but taking the proper precautions can avoid costly loss of analysis time due to software crashes.

A Note on NVivo Server
NVivo manufacturer QSR International has developed a collaborative software solution called NVivo Server. Projects that are stored in NVivo Server can be considerably larger, up to 100 GB. NVivo Server allows multiple users to work on the same project from different computers simultaneously. While useful, in our experience the logistical challenges associated with working on a server have kept our colleagues and clients from using this tool. While we support NVivo server, it is beyond the purposes of this book to describe it. Feel free to follow contact us directly if you and your team have any interest in NVivo Server.

23. MODELS

Models are useful tools when a project is developing or when a project is ready to begin reporting findings. Models present ideas and theories visually. In a research team, models are also useful for team meetings.

Styles for graphical elements that should be used in future projects are created with **Application Options**, under the **Model Styles** tab (see page 43). Styles for graphical elements that should be used in the current project are created with **Project Properties**, under the **Model Styles** tab (see page 56).

Creating a New Model

1. Go to **Explore | Models | New Model**.
   Default folder is **Models**.
   Go to 5.

alternatively

1. Click **[Models]** in Area 1.
2. Select the **Models** folder in Area 2 or its subfolder.
3. Go to **Explore | Models | New Model**.
   Go to 5.

alternatively

3. Click on an empty space in Area 3.
4. Right-click and select **New Model...**

The **New Model** dialog box appears:

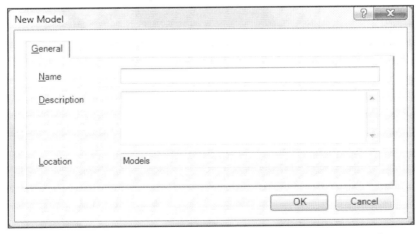

5. Type a name (compulsory) and a description (optional), then **[OK]**.

A new window appears and it is a good idea to undock with **View | Window | Docked** to give you more space on the screen:

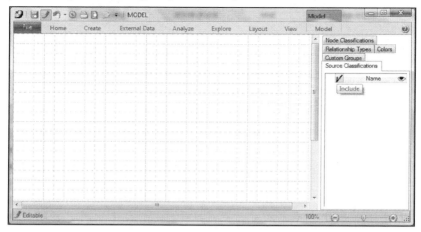

A new Ribbon menu, **Model**, has now opened.

6 Go to **Models | Items | Add Project Items** or right-click and select **Add Project Items...**

The **Select Project Items** dialog appears:

7 Select the **Nodes** folder and then the Node *Foreign Countries*, then [**OK**].

The **Add Associated Data** dialog box is shown. The exact appearance of this dialog box depends on the type of item you have selected in the previous dialog box.

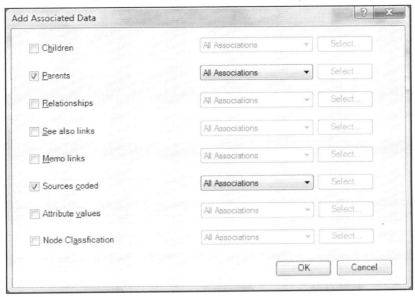

8   We select *Parents* and *Sources Coded*, then [**OK**].

The appearance of the **Add Associated Data** dialog box depends on the type of item. If you select a Source Item, the dialog box looks like this:

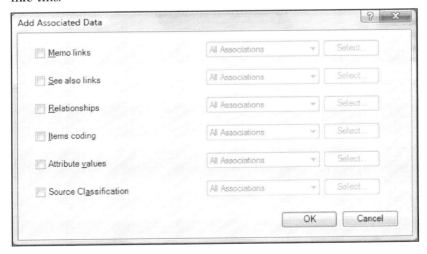

The result may look like this. There are many options to edit the image for added clarity.

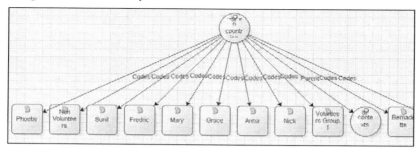

9   Go to **Model | Display | Layout**
    or right-click and select **Layout...**
The **Model Layout** dialog box appears:

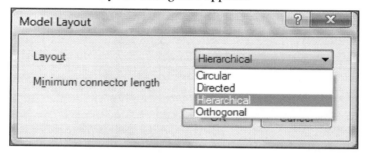

10 Select an option from the **Layout** drop-down list, confirm with **[OK]**.

Display Color Codes

The option to display color codes for individual project items is very useful when it comes to Models:

1   Open a Model in edit mode.
2   Go to **View | Visualization | Color Scheme → Item Colors**.

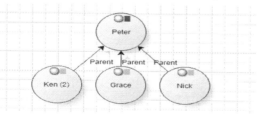

Beside the item symbol a small square is displaying the color code.

300

Creating a Static Model

A static model is a model that is independent of its linked items. A static model cannot be edited.
1. Create or open a dynamic model.
2. Go to **Create | Items | Create As → Create As Static Model...** or right-click and select **Create As → Create As Static Model...**

The **New Model** dialog box is shown.
3. Type a name (compulsory) and a description (optional), then [**OK**].

Creating Model Groups

Model groups allow you to show, hide or select multiple items in an NVivo model.
1. Create or open a dynamic model.

Make sure that the Custom Groups window to the right is shown. Hide and unhide is controlled with **Model | Display | Model Groups**.

2. Click the *Custom Groups* tab in this window.

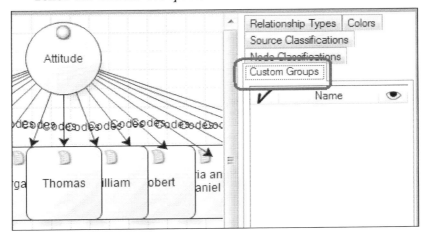

3. Go to **Models | Groups | Group → New Group...**
Click the Custom Groups tab and go to **Model → Group → New Group...**
or click on the Custom Groups window, right-click and select **New Group...**

The **Model Group Properties** dialog box appears:

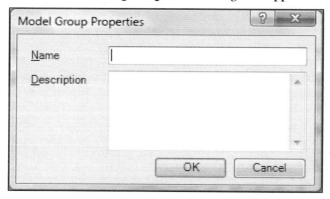

4   Type a name (compulsory) and a description (optional), then [**OK**].
5   Select the graphical items that you want to belong to the new group. Use [**Ctrl**] to select several items.
6   Check in the column marked with ✓ in the row for the new group.

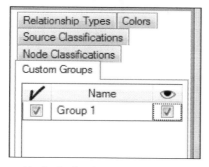

To hide/unhide a certain group, click in the column marked with an eye and in the row for the new group.

Adding More Graphical Shapes

1   Go to **Model | Shapes** → <select> and select a shape from the list
    or position the cursor approximately where you want the shape to be placed, right-click and select **New Shape**. Select a shape from the list.
2   Select the new shape.
3   Go to **Home | Item | Properties**
    or right-click and select **Shape/Connector Properties**.

The **Shape Properties** dialog box appears:

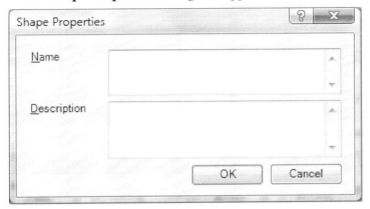

4   In the textbox **Name** type the text that shall be placed in the shape, then [**OK**].

Graphical shapes that have been inserted like this do not have the small symbol that represents a link to a Project Item. When such linked item is deleted a small red cross is placed on the symbol on top of the shape.

Graphical element from an inserted Project Item

Graphical element from an inserted Project Item which has been deleted

Graphical element from an inserted graphical shape

Creating Connectors between Graphical Items

1   Select two graphical items.
2   Go to **Model | Connectors | <select>**
    Select one of these options:

303

Deleting Graphical Items

1. Select one or more graphical items.
2. Use the [**Del**] key
   or go to **Home | Editing → Delete**
   or right-click and select **Delete**.

Converting Graphical Shapes

Graphical shapes can be converted to Linked Project Items:

1. Select a graphical shape.
2. Go to **Model | Items | Convert To → Convert To Existing Project Item**
   or right-click and select **Convert To → Convert To Existing Project Item**.
3. The **Select Project Item** dialog box is shown and you can choose from all existing items except those which are already used in the current model. Used items are dimmed.
4. Confirm with [**OK**].

Linked graphical items can be converted to graphical shapes:

1. Select one or more linked graphical items.
2. Go to **Model | Items | Convert To → Convert To Shape/Connector**
   or right-click and select **Convert To → Convert To Shape/Connector**.

The graphical item retains its shape and only the link is deleted.

Editing a Graphical Item

Editing Associations

When a graphical item has been added it is possible to change or update its associations to other Project Items at any later occasion:

1. Select one or more graphical items.
2. Go to **Model | Items | Add Associated Data**
   or right-click and select **Add Associated Data...**

The **Add Associated Data** dialog box is shown. The content of this dialog box depends on the type of item that you have selected:

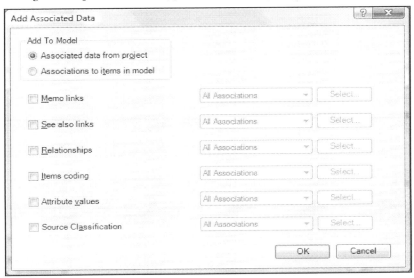

Associated data from project means that data and items from the whole project can be added.

Associations to items in model means that data and items from the current model can be added.

3 When the options that you want have been selected, click **[OK]**.

Changing Text Format
1. Select one or more graphical items.
2. Go to **Home | Format** → <select>.
3. Select font, size and color.

Changing Model Style
Available templates are listed in **Project Properties**, the **Model Styles** tab (see page 56).
1. Select one or more graphical items.
2. Go to **Home | Styles** → <select>.
3. Select a graphical template and confirm with **[OK]**.

Changing Fill Color
1. Select one or more graphical items.
2. Go to **Home | Format | Fill**.

The **Fill** dialog box appears:

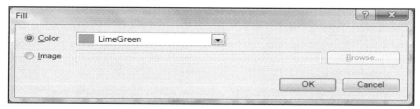

3   Select fill color from the *Color* drop-down list or find an image using the *Image option* and using [**Browse...**].
4   Confirm with [**OK**].

Changing Line Color and Line Style
1   Select one or more graphical items.
2   Go to **Home | Format | Line**.

The **Line** dialog box appears:

3   Select line style, weight, and color.
4   Confirm with [**OK**].

Exporting your Model

NVivo 10 can export your models as either a JPEG, Bitmap, GIF, or SVG image file.

1   Click [**Models**] in Area 1.
2   Select the **Models** folder in Area 2 or its subfolder.
3   Select the Model in Area 3 that you want to export.
4   Go to **External Data | Export | Export → Export Item** or right-click and select **Export → Export Model...** or [**Ctrl**] + [**Shift**] + [**E**].
5   Decide file name, file location, and file type, then [**Save**].

Of interest to researchers who also dabble in graphic design software, such as Adobe InDesign or Microsoft Visio, NVivo allows models to be exported as a Scalable Vector Graphic file format (.svg). These image files are optimized for web viewing and are suitable for importing into graphic design software. After a small amount of tweaking, your NVivo model could be optimized for your next conference poster presentation. Feel free to follow up with us if you want more information on working with your .svg files outside of NVivo 10.

24. MORE ON VISUALIZING YOUR DATA

The visualizations that NVivo offers are:

- Models
- Charts
- Word Trees
- Cluster Analysis
- Tree Maps
- Graphs

Models are dealt with in Chapter 23, Models.

Charts are dealt with in Chapter 12, Coding, section Charts, see page 169.

Word Trees are dealt with in Chapter 13, Queries, section Text Search Queries, see page 186.

Cluster Analysis

Cluster analysis is an exploratory technique that you can use to visualize patterns in your project by grouping sources or Nodes that share similar words, similar attribute values, or are coded similarly by Nodes. Cluster analysis diagrams provide a graphical representation of sources or Nodes to make it easy to see similarities and differences. Sources or Nodes in the cluster analysis diagram that appear close together are more similar than those that are far apart.

1 Go to **Explore | Visualizations | Cluster Analysis**.

The **Cluster Analysis Wizard – Step 1** appears:

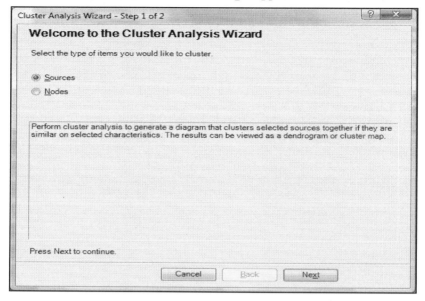

2   We want to analyze selected Source Items, PDF articles. We select the *Sources* options and click [**Next**].

The **Cluster Analysis Wizard – Step 2** appears:

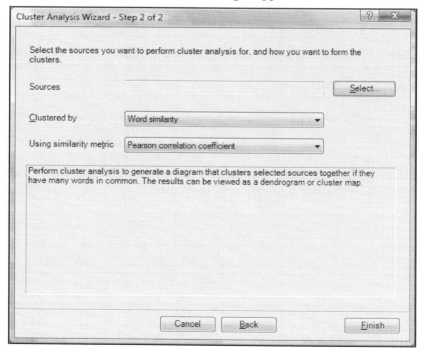

Under the **Clustered by** drop-down list you find the following options: *Word similarity, Coding similarity* and *Attribute value similarity.*

Under the **Using similarity metric** drop-down list you find the following options: *Jaccard's coefficient, Pearson correlation coefficient* and *Sørensen coefficient.*

    3   The [**Select**] button opens the **Select Project Items** dialog box and we select the PDF articles.

    4   Click [**Finish**].

The Ribbon menu **Cluster Analysis** opens and you can choose between 2D Cluster Map, 3D Cluster Map, Horizontal Dendrogram or Vertical Dendrogram. The default diagram is the Horizontal Dendrogram:

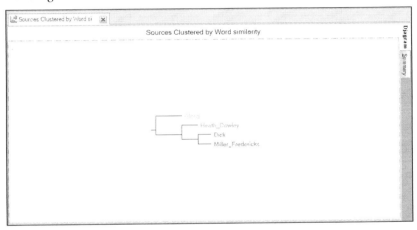

Go to **Cluster Analysis | Type | 2D Cluster Map** and the following diagram is shown:

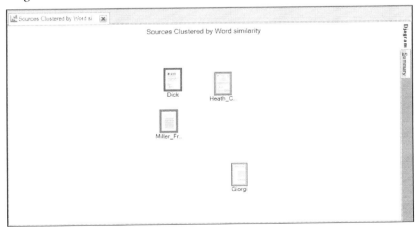

Go to **Cluster Analysis | Type | 3D Cluster Map** and the following diagram is shown:

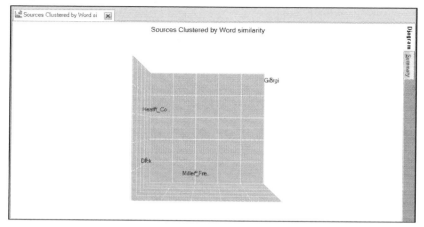

The **Summary** tab to the right shows the current metric coefficient for each pair of items in the cluster:

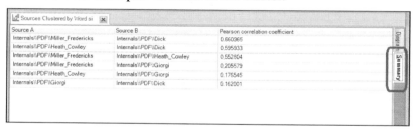

With **Cluster Analysis | Options | Select Data** you can choose the metric coefficient. **Cluster Analysis | Options → Clusters** is any number between 1 and 20 (10 is default) representing the number of colors used in the cluster diagrams.

A Cluster map can also be used like this: Select a word in the graph, right-click and the menu alternatives are: **Open Source** (or double-click or key command **[Ctrl] + [Shift] + [O]**), **Export Diagram, Print, Copy** (the whole graph), **Run Word Frequency Query, Item Properties, Select Data**.

Tree Maps

Tree Maps are a way to demonstrate how a Source Items or Nodes are related to selected information:
1 Go to **Explore | Visualizations | Tree Maps**.
The **Tree Map Wizard – Step 1** appears:

We want to analyze our interviews.
2 Click **[Next]**.
The **Tree Map Wizard – Step 2** appears:

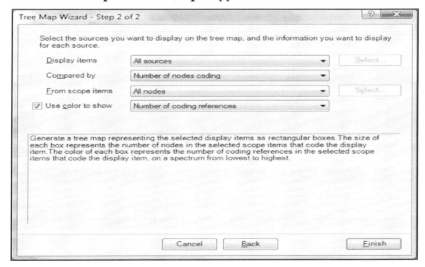

3   Click [**Finish**].

The Ribbon menu **Tree Map** appears and the result may look like this:

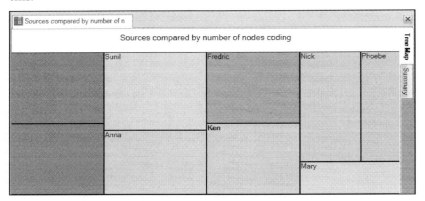

The **Summary** tab to the right shows the current number of coding references and number of Nodes coding:

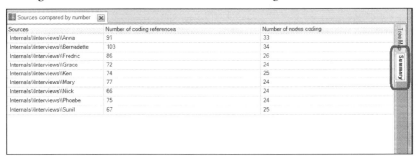

With **Tree Map | Options | Color Scheme** you can change between a color scheme for number of Nodes Coding or Number of Coding References or Item Colors. You can also choose from four different Color Spectra. **Tree Map | Options | Select Data** you will bring forward a dialog box identical with **Tree Map Wizard – Step 2** and you can modify the Tree Map.

A Tree Map can also be used like this: Select a word in the graph, right-click and the menu alternatives are: **Open Source** (or double-click or key command [**Ctrl**] + [**Shift**] + [**O**]), **Export Diagram, Print, Copy** (the whole graph), **Item Properties, Select Data**.

Graphs

Graphs are a quick and simple method to demonstrate how a selected Source Item or Node is related to other items:
1. Select the single item in Area 3 that you want to analyze.
2. Go to **Explore | Visualizations | Graph**.

The Ribbon menu **Graph** appears and the following image appears directly in Area 4:

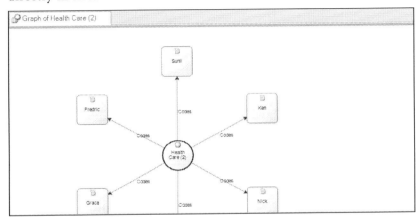

With **Graph | Display <options>** you can select which associated data you want to show in the graph.

The graph has the same design as Models, see Chapter 23, Models, and can also be saved as a 'genuine' model by going to **Graph | Create | Create Model from Graph**.

A Graph can also be used like this: Select an item in the graph, right-click and the menu alternatives are: **Open Item** (or key command **[Ctrl] + [Shift] + [O]**), **Graph** (creates a new graph), **Item Properties** (or double-click), **Export Graph, Print, Copy** (the whole graph), **Create Model from Graph**.

25. REPORTS AND EXTRACTS

Reports contain summary information about your project that you can view and print. For example, you could check the progress of your coding by running a report that lists your sources and the Nodes that code them.

An extract lets you export a collection of data to a text, Excel or XML file. In some cases you can use this data for complementary analysis in other applications.

Understanding Views and Fields

In reports and extracts, a view is a group of related data fields. There are five different views: Source, Source Classification, Node, Node Classification, and Project Items. When you build a report or an extract, select the view which contains the fields you want to include in your report or extract.

View	Comment
Source	Report on sources including which Nodes code the sources. This view also includes collections, which you could use to limit the scope of your reports.
Source Classification	Report on the classifications that are used to describe your sources. You can create reports that show the classifications in your project or how your sources are classified. This view does not contain any coding information. To report on coding in sources, choose the Source view.
Node	Report on the Nodes in your project including sources they code, coding references, and any classifications assigned to them. This view includes 'intersecting' Nodes which is useful for reporting on how coding at two Nodes coincides—for example see which 'cases' intersect selected themes. This view also includes collections, which you could use to limit the scope of your reports.
Node Classification	Report on the classifications, attributes and attribute values used to describe the people, places and other cases in your project. You can use this view to show the classification structure, or the demographic spread of classified Nodes. This view does not contain any coding information. If you want to report on coding at Nodes, choose the Node view.
Project Items	Use this view to create reports about the structure of your project. Report on your project and the Project Items, including the types of Project Items and who created them.

Report and Extract Templates

NVivo comes with 8 pre-defined Report templates and 8 pre-defined Extract templates ready to be used for any NVivo project. These templates can be deleted or modified by the user. New Reports and Extracts can be created by the user for the current project or be exported to other users working with any other NVivo projects. If you want to inherit Reports and Extracts from one project to the next, then check *Add predefined reports/extracts to new projects*, in the **Application Options** dialog box, see page 36.

The pre-defined Report templates are located in the **Reports** folder under the **[Reports]** navigation button:

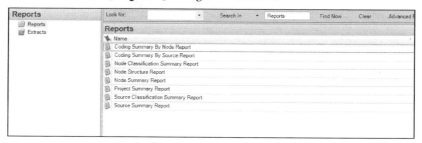

The Report templates that include contents are 'Coding Summary By Node Report' and 'Coding Summary By Source Report'.

The pre-defined Extract templates are located in the **Extracts** folder under the **[Reports]** navigation button:

The Extract templates that include contents are 'Coding Summary By Node Extract' and 'Coding Summary By Source Extract'.

Reports

Creating a New Report via the Report Designer

1. Go to **Explore | Reports | New Report → New Report via Designer...**

The **New Report** dialog box appears:

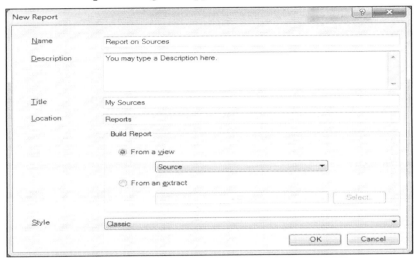

A new Ribbon menu, **Report**, now appears. If you prefer the option **From an extract** you will need to select an extract from the existing extract templates and you will inherit the view and fields from the chosen extract. We have typed a name and title of the Report and we have selected the View *Source*.

2. Click **[OK]**.

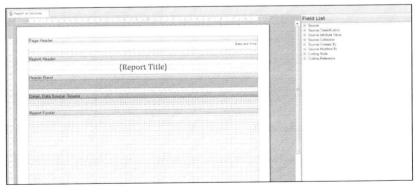

The key to understanding the **Report Designer** is the idea of controls. Controls are static fields with label data or dynamic fields with data. The following graph will explain. First you select a field

317

from the list of field headings in the Field List panel to the right. Then go to **Report | Add/Modify** or right-click and select **Add Field**. The result is two controls, Label Control and Field Control.

The text or Image controls are created when you click on an empty space immediately below one of the band headers, for example the Report Header. Then an empty control, which is rectangular, is created. From here you go to **Report | Header & Footer** for insertion of Report Title, Report Location, User Name, Date and Time, Project Name or Page N of M.

You can also create your own text box or an image like a logo. Go to **Reports | Controls** and choose **Text** or **Image**. All such controls can easily be resized, moved, deleted etc. Editing text is made by double-clicking the text in question and then either the **Modify Text** or **Modify Label** dialog box is shown.

In case you need to edit fonts, color or size then select the control and go to **Home | Format** and make the modifications you need.

Report | Page | Layout can be used to go between Tabular or Columnar layout for a non-grouped report and Stepped, Blocked or Outlined layout for a grouped report.

A Grouped report is created by going to **Report | Grouping | Group** and then you select the field or fields that will create structural headings in the report.

Report | Sort & Filter | Sort makes it possible to change the sorting principles and **Report | Sort & Filter | Filter** offers an option to introduce a filter either with fixed or user-prompted settings.

Example of a Report based on Source View

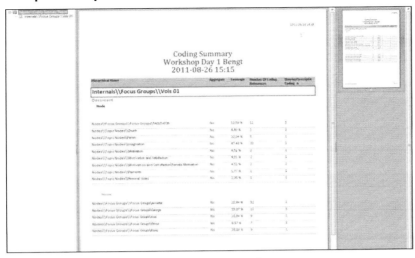

Creating a New Report via the Report Wizard

The Report Wizard provides a systematic method for creating a new, custom report for your project:

1. Go to **Explore | Reports | New Report → New Report via Wizard...**

The **Report Wizard – Step 1** appears:

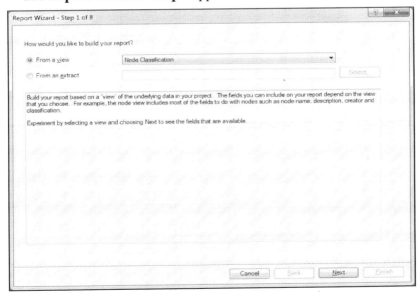

2. Select *Node Classification* from the **From a view** drop down list. If you prefer the option **From an extract** you will need to select an extract from the existing extract

templates and you will inherit the view and fields from the chosen extract.

3   Click [**Next**].

The **Report Wizard – Step 2** appears:

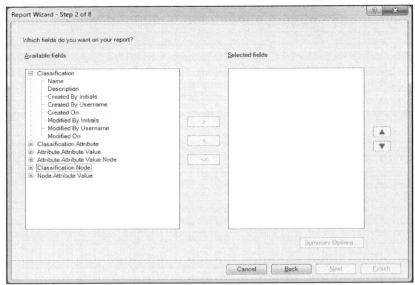

4   Expand the field headings and select the fields from the left box that will form the Report and click the [>] button which brings over the fields to the right box.

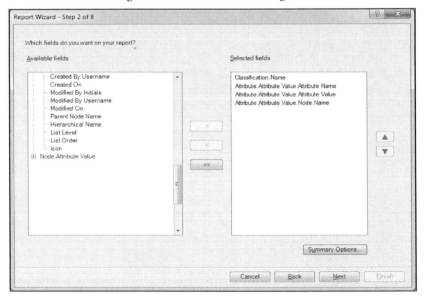

5   Click [**Next**].

The **Report Wizard – Step 3** appears:

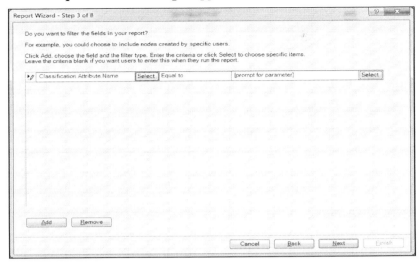

6   Use the **[Add]** button to create the first filter row and then the **[Select]** button to select the field that shall limit the report. If you leave the right textbox as *[prompt for parameter]* then the user will be prompted to select a parameter each time the report is run.
7   Click **[Next]**.

The **Report Wizard – Step 4** appears:

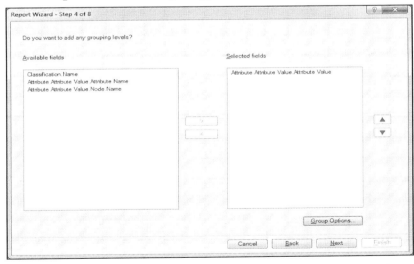

8   Grouping is a way to introduce headings in the Report thus making the report easier to read. We select *Attribute.Attribute Value.Attribute Value* and use the [>] button to bring it over to the right box.
9   Click **[Next]**.

The **Report Wizard – Step 5** appears:

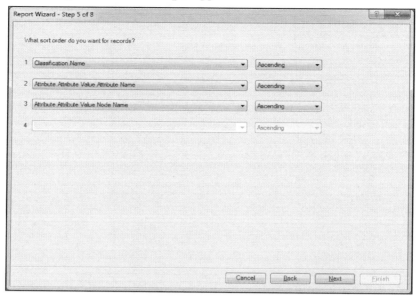

10   We decide the sort order by using the drop-down lists.
11   Click [**Next**].

The **Report Wizard – Step 6** appears:

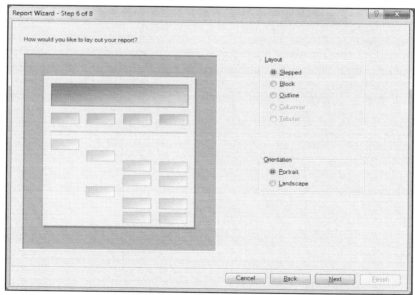

12   We accept the default settings which are *Stepped* layout and *Portrait* orientation.
13   Click [**Next**].

The **Report Wizard** – **Step 7** appears:

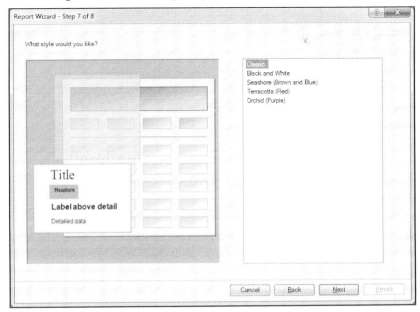

14 We accept the default setting which is the *Classic* style.
15 Click **[Next]**.

The **Report Wizard** – **Step 8** appears:

16 Finally, we type the Name and Title of the Report and optionally a Description.

17 Click [**Finish**].
The new **Report** opens:

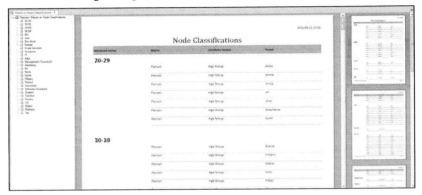

The left panel is called **Report Map** and can be used to easily find a certain headings in the Report. The right panel is called **Thumbnails** and can be used to find a certain page. Report Map and Thumbnails can be hidden/unhidden with **View | Detail View | Report** and **Report Map** or **Thumbnails**, two toggling functions.

From this view you can print the Report or export the Report as a Word document.

Exporting the Result of a Report

1. Click [**Reports**] in Area 1.
2. Select the **Reports** folder in Area 2.
3. Select a report in Area 3.
4. Open the Report.
5. Go to **Export | Export → Export Report Results** or [**Ctrl**] + [**Shift**] + [**E**].
6. Decide file name, file type (Word, text formats, Excel formats, PDF, RTF, Web formats) and location. Click [**Save**].

Exporting a Report Template

1. Click [**Reports**] in Area 1.
2. Select the **Reports** folder in Area 2.
3. Select a report in Area 3.
4. Go to **Export | Export → Export Report** or right-click and select **Export → Export Report** or [**Ctrl**] + [**Shift**] + [**E**].
5. Decide file name and location. The file type is already determined as .NVR. Click [**Save**].

The result is a Report template that can be imported and used by other projects.

Importing a Report Template
1. Go to **External Data | Import | Report**.
   Default folder is **Reports**.
   Go to 5.

alternatively
1. Click on [**Reports**] in Area 1.
2. Select the **Reports** folder in Area 2 or its subfolder.
3. Go to **External Data | Import | Report**.
   Go to 5.

alternatively
3. Click on any empty space in Area 3.
4. Rightclick and select **Import Report...**
5. The **Import Report** dialog box appears.
6. Select the Report template .NVR that you want to import.
   Click on [**Open**].

The **Report Properties** dialog box appears:

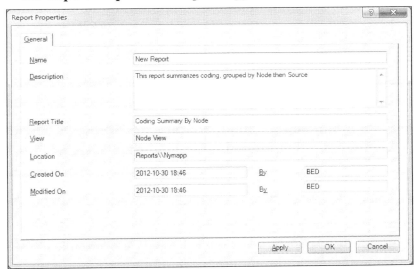

7. If you need you can change or modify the text in the dialog box. Click on [**OK**].

Editing a Report
By opening a Report in Report Designer you can modify any parameter except the selected View and Style:
1. Click [**Reports**] in Area 1.
2. Select the **Reports** folder in Area 2.
3. Select a report in Area 3.
4. Go to **Home | Item | Open → Open Report in Designer...**
   or right-click and select **Open Report in Designer...**
   or [**Ctrl**] + [**Shift**] + [**O**].

The result may look like this:

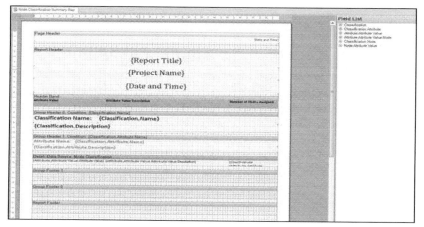

Here you can make any modification that Report Designer allows.

The ribbon tab **Reports** opens. Here you can change the layout, modify filters, modify headers and footers, change grouping and sorting. You can also add your own text and/or your own logo.

Extracts

Creating a New Extract

An extract lets you export a portion of your data to a text file, an Excel spreadsheet, or an XML file.

1 Go to **Explore | Reports | New Extract...**

The **Extract Wizard - Step 1** appears:

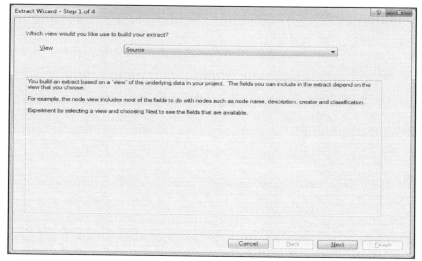

2 Select *Source* from the View drop down list.
3 Click [**Next**].

The **Extract Wizard** – **Step 2** appears:

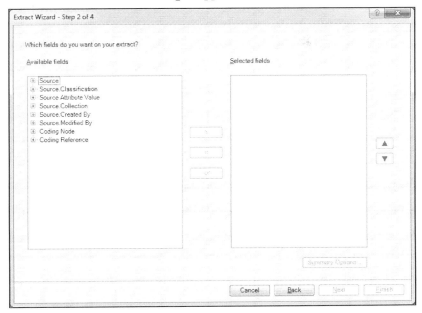

4   Expand the field headings and select the fields from the left box that will form the Extract and click the [>] button which brings over the fields to the right box.

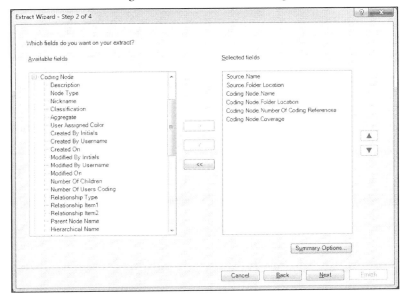

5   Click [**Next**].

The **Extract Wizard – Step 3** appears:

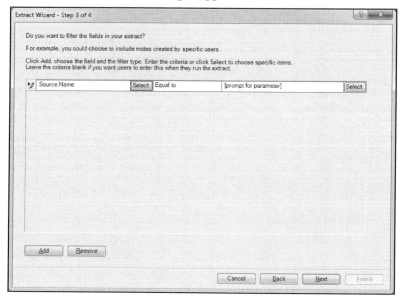

6   Use the [**Add**] button to create the first filter row and then the [**Select**] button to select the field that shall limit the report. Leave the right textbox as *[prompt for parameter]* and the user will be prompted to select a parameter each time the extract is run.

7   Click [**Next**].

The **Extract Wizard** – **Step 4** appears:

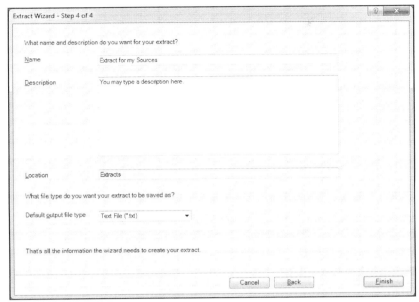

8 Type a name of the Extract (compulsory) and optionally a description. Default file format can also be set but can be changed before you run the Extract. The file formats that you can choose from are: Text, Excel, and XML.
9 Confirm with **[Finish]**.

Exporting (Running) an Extract
1 Click **[Reports]** in Area 1.
2 Select the **Extracts** folder in Area 2.
3 Select an extract in Area 3.
4 Go to **Reports | Run Extract**
   or double-click
   or right-click and select **Run Extract**.
5 Decide file name, file type and location. Click **[Save]**.

Exporting an Extract Template
1 Click **[Reports]** in Area 1.
2 Select the **Extracts** folder in Area 2.
3 Select an extract in Area 3.
4 Go to **External Data | Export | Export → Export Extract**
   or right-click and select **Export → Export Extract**
   or **[Ctrl]** + **[Shift]** + **[E]**.
Decide file name and location. The file type is already determined as .NVX. The result is an Extract template that can be imported and used by other projects.

Importing an Extract Template

1   Go to **External Data | Import | Extract**.
    Default folder is **Extracts**.
    Gå till 5.

alternatively

1   Click on **[Reports]** in Area 1.
2   Select **Extracts** folder in Area 2 or its subfolder.
3   Go to **External Data | Import | Extract**.
    Go to 5.

alternatively

3   Click on any empty space in Area 3.
4   right-click and select **Import Extract...**
5   The **Import Extract** dialog box appears.
6   Select the Extract template .NVX that you want to import.
    Click on **[Open]**.

The **Extract Properties** dialog box appears:

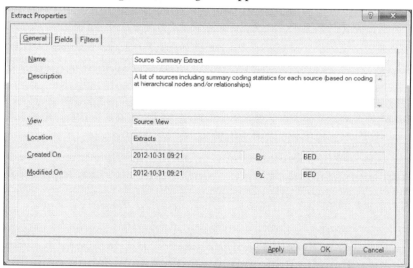

7   If you need you can change or modify the text in the dialog box. Click on **[OK]**.

Editing an Extract

An Extract can be modified by opening its Extract Properties:

1   Click **[Reports]** in Area 1.
2   Select **Extracts** folder in Area 2.
3   Select an extract in Area 3.
4   Go to **Home | Item | Properties → Extract Properties...**
    or right-click and select **Extract Properties**
    or **[Ctrl] + [Shift] + [P]**.

26. HELP FUNCTIONS IN NVIVO

An integral part of NVivo is the variety of help and support functionality for users. You can choose from Online Help or Offline Help. Changes are made in the **File → Options** dialog box:

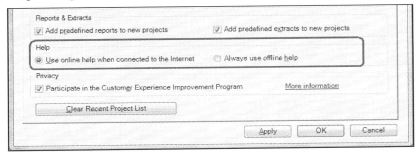

Help Documents

1. Go to **File → Help → NVivo Help**
   or use the [?] symbol in the upper right corner of the screen or [F1].

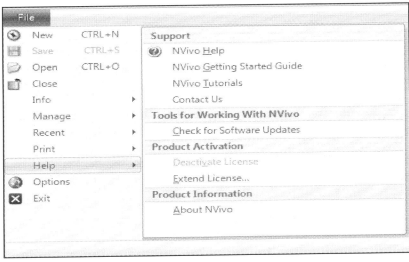

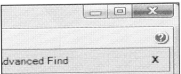

The initial view for **Oneline Help** is this:

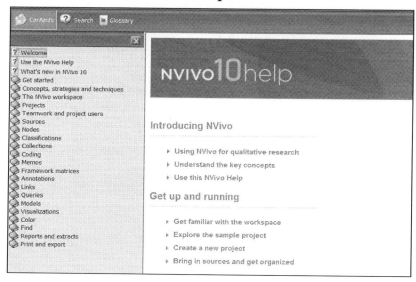

Tutorials

NVivo has some tutorials in the form of video clips:
1. Go to **File → Help → NVivo Tutorials**.

Users can also access QSR's online tutorials. Adobe Flash Player is required to play these tutorials.

Support and Technical Issues

As a holder of this book you are welcome to contact **support@formkunskap.com** or Skype **bengt.edhlund** in any matter that has to do with installation problems or user procedures as described in this book.

In case of performance disturbances, an error log is created automatically. The log files are by default stored in My Documents folder of the current user. Such error log file has the following name structure '**err**<**date**>**T**<**time**>**.log**'. It is a text file and in case you need technical assistance you may be asked to forward such error log file to QSR Support or the local representative for analysis.

Software Versions and Service Packs

You should always be aware of the software version and Service Pack that you use. A Service Pack is an additional software patch that could carry bug fixes, improvements and new features. Service Packs are free for licensees of a certain software version. Provided that you are connected to the Internet and have enabled *Check for Update every 7 Days* (see page 38) you will automatically get a message on the screen when a new Service Pack has been launched. Always use the latest available Service Pack:

1. Go to **File → Help → About NVivo**.

The image shows the software version and installed Service Pack:

27. GLOSSARY

This is a list of the most common words, terms, and descriptions that are used in this book.

Advanced Find	Search names of Project items like Source items, Memos or Nodes. Use **Find Bar** - **Advanced Find.**
Aggregate	Aggregate means that a certain Node in any hierarchical level accumulates the logical sum of all its nearest Child Nodes.
Annotation	A note linked to an element of a Source item. Similar to a conventional footnote.
Attribute	A variable that is used to describe individual Source items and Nodes. Example: age group, gender, education.
Autocoding	An automatic method to code documents using names of the paragraph styles.
Boolean Operator	The conventional operators AND, OR or NOT used to create logical search expressions applying Boolean algebra.
Case Node	A Case Node is a member of a group of Nodes which are classified with Attributes and Values reflecting demographic or descriptive data. Case Nodes can be people (Interviewees), places or any group of items with similar properties.
Casebook	Definition used in NVivo 8 corresponding to Classification Sheet in later versions of NVivo.
Classification	A collection of Attributes for Source items or Nodes.
Classification Sheet	A matrix overview of the attributes and values of Source items or Nodes.
Cluster Analysis	Cluster analysis or clustering is the assignment of a set of observations into subsets (called *clusters*) so that observations in the same cluster are similar in some sense. Clustering is a method of unsupervised learning, and a common technique for statistical data analysis used in many fields, including machine learning, data mining, pattern recognition, image analysis, information retrieval, and bioinformatics.

Coding	The work that associates a certain element of a Source item at a certain Node.
Coding Stripe	Graphical representation of coding in a Source item.
Coding Queries	A method to construct a query by using combinations of Nodes or Attribute values.
Compound Queries	A method to construct a query by using combinations of various query types.
Coverage	The fraction of a Source item that has been coded at a certain Node.
Dataset	A structured matrix of data arranged in rows and columns. Datasets can be created from imported Excel spreadsheets or captures social media data.
Dendrogram	A tree-like plot where each step of hierarchical clustering is represented as a fusion of two branches of the tree into a single one. The branches represent clusters obtained at each step of hierarchical clustering.
Discourse Analysis	In semantics, discourses are linguistic units composed of several sentences — in other words, conversations, arguments or speeches. Discourse Analysis studies how texts can be structured and how its elements are interrelated.
Document	An item in NVivo that is usually imported from a Source document.
Dropbox	A cloud-based software solution that allows file syncing across several computers.
EndNote	A powerful and convenient reference handling software tool.
Ethnography	The science that examines characteristics of different cultural groups.
Evernote	A popular cloud-based notetaking platform that creates text and voice memos.
Facebook	A social networking platform where users can become 'friends' and post content on one another's personal page ('walls'). Social groups are also available in Facebook (pages).

Filter	A function that limits a selection of values or items in order to facilitate the analysis of large amounts of data.
Find Bar	A toolbar immediately above the List View.
Focus Group	A selected, limited group of people that represents a larger population.
Folder	A folder that is created by NVivo is a virtual folder but has properties and functions largely like a normal Windows folder.
Framework	A data matrix that allows you to easily view and summarize areas of your data you wish to more closely explore.
Grounded Theory	Widely recognized method for qualitative studies where theories emerge from data rather than a pre-determined hypothesis.
Grouped Find	A function for finding items that have certain relations to each other.
Hushtag	A 'keywording' convention that places a number sign (#) before a term in order to allow text-based searches to distinguish searchable keywords from standard discourse (see also, Twitter).
Hyperlink	A link to an item outside the NVivo-project. The linked item can be a file or a web site.
In Vivo Coding	In Vivo coding is creating a new Node when selecting text and then using the *In Vivo* command. The Node name will become the selected text (max 256 characters) but the name (and location) can be changed later.
Items	All items that constitutes a project. Items are Sources, Nodes, Classifications, Queries, Results, and Models.
Jaccard's Coefficient	The **Jaccard index**, also known as the **Jaccard similarity coefficient** (originally coined *coefficient de communauté* by Paul Jaccard), is a statistic used for comparing the similarity and diversity of sample sets

Kappa Coefficient	**Cohen's kappa coefficient, (K),** is a statistical measure of inter-rater agreement. It is generally thought to be a more robust measure than simple percent agreement calculation since **K** takes into account the agreement occurring by chance. Cohen's kappa coefficient measures the agreement between two raters who each classify N items into C mutually exclusive categories. If the raters are in complete agreement then **K** = 1. If there is no agreement among the raters (other than what would be expected by chance) then **K** $\leq$ 0.	
LinkedIn	A professional social networking site where users become 'connections' and participate in group discussions in 'groups'.	
Matrix Coding Query	The method to construct queries in a matrix form where contents in each cell are the result of a row and a column combined with a certain operator.	
Medline	The world's most popular health research database.	
Memo Link	Only *one* Memo Link can exist from an item to a memo.	
Memo	A text document that could be linked from *one* Source item or from *one* Node.	
MeSH	MeSH (Medical Subject Headings), the terminology or controlled vocabulary used in PubMed and associated information sources.	
Mixed Methods	A combination of quantitative and qualitative studies.	
Model	Graphical representation of Project items and their relations.	
Node	Often used in the context of a 'container' of selected topics or themes. A Node contains pointers to whole documents or selected elements of documents relevant to the specific Node. Nodes can be organized hierarchically.	
OCR	Optical Character Recognition, a method together with scanning making it possible to identify characters not only as an image.	
OneNote	Microsoft's cloud-based notetaking platform that creates text and voice memos.	

Pearson Correlation Coefficient	A type of correlation coefficient that represents the relationship between two variables that are measured on the same interval or ratio scale.
Phenomenology	A method which is descriptive, thoughtful, and innovative and from which you might verify your hypothesis.
Project	The collective denomination of all data and related work.
PubMed	A popular health research database (cf. Medline).
Qualitative Research	Research with data originating from observations, interviews, and dialogs that focuses on the views, experiences, values, and interpretations of participants.
Quantitative Research	Research that colletcs data through measurements and conclusions through calculations and statistics.
Ranking	The organization of results according to ascending or descending relevance.
RefWorks	A popular reference handling software tool.
Relationship	A Node that defines a relation between two Project items. A relationship is always characterized by a certain relationship type.
Relationship Type	A concept (often a descriptive verb) that defines a relationship or dependence between two Project items.
Relevance	Relevance in a result of a query is a measure of success or grade of matching. Relevance may be calculated as the number of hits in selected sections of the searched item.
Research Design	A plan for the collection and study of data so that the desired information is reached with sufficient reliability and a given theory can be verified or rejected in a recognized manner.
Result	A result is the answer to a query. A result may be shown as *Preview* or saved as a *Node*.
Saving Queries	The possibility to save queries in order to re-run or to modify them.
See Also Link	A link established between two items. A See Also Link is created from a certain area or text element of an item to a selected area or the whole of another item.

Service Pack	Software updates that normally carry bug fixes, performance enhancements, and new features.
Set	A subset or 'collection' of selected Project Items. A saved set can be displayed as a list of shortcuts to these Project Items.
Sørensen Coefficient	The **Sørensen index**, also known as **Sørensen's similarity coefficient**, is a statistic used for comparing the similarity of two samples. It was developed by the botanist Thorvald Sørensen and published in 1948.
Stop Words	Stop words are less significant words like conjunctions or prepositions, that may not be meaningful to your analysis. Stop words are exempted from Text Search Queries or Word Frequency Queries.
Twitter	A social networking website where users post 'tweets' that contain a maximum of 140 characters.
Uncoding	The work that deletes a given coding of a document at a certain Node.
Validity	The validity of causal inferences within scientific studies, usually based on experiments.
Value	Value that a certain Attribute can have. Similar to 'Controlled Vocabulary'. Example: male, female.
Zotero	A reference handling software tool.

APPENDIX A – THE NVIVO SCREEN

The NVivo Screen

1. The Navigation Buttons
2. The Virtual Explorer
3. The List View
4. The Details – Opened Items

The NVivo Screen is similar to that of Microsoft Outlook. Normally you start with the Navigation Buttons (**1**) and select a certain group of folders (**2**). Clicking a folder lists its contents of documents or items (**3**). An item is opened with a double-click and is shown in area (**4**). This window can also be undocked.

For continued work you can either rightclick with the mouse (depending on its position), use Ribbon menus, or key board commands.

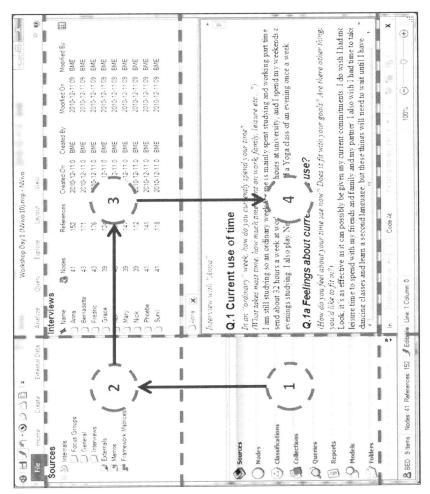

APPENDIX B – KEYBOARD COMMANDS

Listed below are some of the most useful keyboard commands. Many of these strictly adhere to general Windows conventions. Others are specific for each program.

Windows	Word	NVivo 10	Keyboard Command	Desription
✓	✓	✓	[Ctrl] + [C]	Copy
✓	✓	✓	[Ctrl] + [X]	Cut
✓	✓	✓	[Ctrl] + [V]	Paste
✓	✓	✓	[Ctrl] + [A]	Select All
✓	✓	✓	[Ctrl] + [O]	Open Project
	✓[14]	✓	[Ctrl] + [B]	Bold
	✓[14]	✓	[Ctrl] + [I]	Italic
	✓[14]	✓	[Ctrl] + [U]	Underline
	✓[14]		[Ctrl] + [K]	Insert Hyperlink
		✓	[Ctrl] + [E]	Switch between Edit mode and Read Only
		✓	[Ctrl]+[Shift]+[K]	Link to New Memo
		✓	[Ctrl]+[Shift]+[M]	Open Linked Memo
		✓	[Ctrl]+[Shift]+[N]	New Folder/Item
		✓	[Ctrl]+[Shift]+[P]	Folder/Item Properties
		✓	[Ctrl]+[Shift]+[O]	Open Item
		✓	[Ctrl]+[Shift]+[I]	Import Item
		✓	[Ctrl]+[Shift]+[E]	Export Item
		✓	[Ctrl]+[Shift]+[F]	Advanced Find
		✓	[Ctrl]+[Shift]+[G]	Grouped Find
		✓	[Ctrl]+[Shift]+[U]	Move Up
		✓	[Ctrl]+[Shift]+[D]	Move Down
		✓	[Ctrl]+[Shift]+[L]	Move Left
		✓	[Ctrl]+[Shift]+[R]	Move Right
		✓	[Ctrl]+[Shift]+[T]	Insert Time/Date

[14] Only for English version of Word.

Windows	Word	NVivo 10	Keyboard Command	Description
	✓	✓	[Ctrl] + [G]	Go to
✓	✓	✓	[Ctrl] + [N]	New Project
✓	✓	✓	[Ctrl] + [P]	Print
✓	✓	✓	[Ctrl] + [S]	Save
		✓	[Ctrl] + [M]	Merge Into Selected Node
		✓	[Ctrl] + [1]	Go Sources
		✓	[Ctrl] + [2]	Go Nodes
		✓	[Ctrl] + [3]	Go Classifications
		✓	[Ctrl] + [4]	Go Collections
		✓	[Ctrl] + [5]	Go Queries
		✓	[Ctrl] + [6]	Go Reports
		✓	[Ctrl] + [7]	Go Models
		✓	[Ctrl] + [8]	Go Folders
✓	✓		[Ctrl] + [W]	Close Window
✓	✓		[Ctrl]+[Shift]+[W]	Close all Windows of same Type
	✓	✓	[F1]	Open Online Help
		✓	[F4]	Play/Pause
		✓	[F5]	Refresh
	✓	✓	[F7]	Spell Check
		✓	[F8]	Stop
		✓	[F9]	Skip Back
		✓	[F10]	Skip Forward
		✓	[F11]	Start Selection
		✓	[F12]	Finish Selection
	✓	✓	[Ctrl] + [Z]	Undo
		✓	[Ctrl] + [Y]	Redo
	✓	✓	[Ctrl] + [F]	Find
	✓	✓	[Ctrl] + [H]	Replace (Detail View)
		✓	[Ctrl] + [H]	Handtool (Print Preview)
		✓	[Ctrl] + [Q]	Go to Quick Coding Bar

Windows	Word	NVivo 10	Keyboard Command	Description
		✓	[Ctrl]+[Shift]+[F2]	Uncode Selection at Existing Nodes
		✓	[Ctrl]+[Shift]+[F3]	Uncode Selection at This Node
		✓	[Ctrl]+[Shift]+[F5]	Uncode Sources at Existing Nodes
		✓	[Ctrl]+[Shift]+[F9]	Uncode Selection at Nodes visible in Quick Coding Bar
		✓	[Ctrl] + [F2]	Code Selection at Existing Node
		✓	[Ctrl] + [F3]	Code Selection at New Node
✓	✓	✓	[Ctrl] + [F4]	Close Current Window
		✓	[Ctrl] + [F5]	Code Sources at Existing Node
		✓	[Ctrl] + [F6]	Code Sources at New Node
		✓	[Ctrl] + [F8]	Code In Vivo
		✓	[Ctrl] + [F9]	Code Selection at Nodes visible in Quick Coding Bar
		✓	[Alt] + [F1]	Hide/Show Navigation View
		✓	[Ctrl] + [Ins]	Insert Row
		✓	[Ctrl] + [Del]	Delete Selected Items in a Model
		✓	[Ctrl]+[Shift]+[T]	Insert Date/Time
		✓	[Ctrl]+[Shift]+[Y]	Insert Symbol
✓			[Ctrl]+[Alt]+[F]	Insert Footnote
✓	✓		[Ctrl] + [Enter]	Insert Page break
		✓	[Ctrl] + [Enter]	Carriage Return in certain text boxes

INDEX

A

Aggregate, 129, 335
Annotation, 335
Annotations, 125
Apply, 214
Attributes, 139
audio formats, 91
Auto Scroll, 236
Auto Summary, 231
Autocode, 158, 249, 335
Automatically select hierarchy, 209
Automatically select subfolders, 209

B

bar diagram, 172
Boolean Operator, 335

C

Case Node, 335
Charts, 169
Classifications, 139
Classifying, 241
Cluster Analysis, 181, 308, 335
Cluster Map, 181, 309
Codable, 241
Code sources at new cases located under, 64, 93, 108, 119
Coding, 155, 156
Coding Comparison Queries, 290
Coding Context, 165
Coding Density Bar, 167, 168
Coding Excerpt, 162, 164
Coding Queries, 187
Coding Stripes, 167
  Sub-Stripes, 293
color marking, 23
column diagram, 172
Compound Queries, 193
Connection Map, 208

Context Words, 186
Copy, 28, 204
Copy Project, 62
Copyright, 2
Coverage, 162, 336
Create descriptions, 64, 86, 93, 108, 119
Create Results as New Node, 211
Create results if empty, 211
creating
  a Child Node, 132
  a Classification, 140
  a Document, 65
  a folder, 21
  a Framework Matrix, 229
  a Media Item, 94
  a Memo Link, 121
  a Memo Link and a Memo, 121
  a Model, 297
  a Node, 131
  a Picture Log, 111
  a Relationship, 135
  a Relationship Type, 134
  a Report, 319
  a See Also Link, 122, 124
  a Set, 29
  a Static Model, 301
  a table, 77
  an Annotation, 125
  an Attribute, 142
  an Hyperlink, 126
  Model Groups, 301
  subfolders, 21
Current Node(s), 153
current user, 287

D

Datasets, 241
deleting
  a Folder, 22
  a Framework Matrix, 237
  a Hyperlink, 127

a Memo Link, 122
a See Also Link, 125
an Annotation, 126
an Item, 26
graphical items, 304
dendrogram, 181, 309, 336
Description, 52, 64, 86, 93, 108
Detail View, 47
dialog box
- Add Associated Data, 299
- Advanced Find, 210, 280
- Application Options, 36
- Attribute Properties, 143
- Audio Properties, 23, 93
- Auto Code, 159
- Chart Options, 173
- Classification Filter Options, 146
- Coding Comparison Query, 291
- Coding Query, 188, 190, 191, 192
- Coding Query Properties, 214
- Coding Search Item, 190, 191, 198
- Compound Query, 194
- Dataset Properties, 247
- Delete Confirmation, 122
- Document Properties, 65
- Export Classification Sheets, 148
- Export for NVivo, 275
- Export Options, 61, 67, 70, 106, 115, 248
- Export Project Data, 61
- External Properties, 70
- Fill, 306
- Find Content, 73
- Find Project Items, 209
- Font, 72
- Import from Evernote, 272
- Import from OneNote, 276
- Import Internals, 63, 92, 107
- Import Memos, 118
- Import Project, 60
- Import Transcript Entries, 102
- Insert Text Table, 77
- Line, 306
- Mapping and Grouping Options, 256
- Matrix Coding Query, 197
- Matrix Filter Options, 202
- Merge Into Node, 133
- Model Group Properties, 302
- Model Layout, 300
- New Attribute, 142
- New Audio, 95
- New Classification, 141
- New Document, 66
- New External, 68
- New Folder, 22
- New Framework Matrix, 229
- New Hyperlink, 126
- New Memo, 120
- New Model, 297
- New Node, 131, 132, 134
- New Relationship Type, 135
- New Report, 317
- New Search Folder, 284
- New See Also Link, 123
- New Set, 30
- New Video, 95
- Node Properties, 144
- Page Setup, 82
- Paste, 205
- Paste Special Options, 29
- Picture Properties, 109
- Print Options, 79, 81
- Range Code, 160
- Replace Content, 74
- Save As, 204
- Save Reminder, 62
- See Also Link Properties, 117
- Select Location, 185
- Select Project Item, 121
- Select Project Items, 188, 298
- Select Set, 30
- Shape Properties, 303
- Spelling <Language>, 75
- Subquery Properties, 194
- Text Search Query, 210

Video Properties, 93
Word Frequency Query, 178
Zoom, 78
Discourse Analysis, 336
Dock, 27
documents, 63
double-click, 76
Drag-and-Drop, 132, 153, 155

E

editing
  a Graphical Item, 304
  a Query, 213
  a Report, 325
  an Extract, 330
  Pictures, 112
  text, 71, 83
EndNote, 219, 336
error log file, 332
Ethnography, 336
Evernote, 271, 275, 336
Export List, 26
Export Options, 61, 67, 106, 115, 248
exporting
  a Dataset, 248
  a list, 26
  a Picture Item, 115
  a Report, 324
  a Report Template, 324
  an External Item, 70
  an Extract, 329
  an Extract Template, 329
  Audio/Transcript, 105
  Classification Sheets, 147
  Documents, 67, 237
  Framework Matrices, 237
  Project Data, 60
  Video/Transcript, 105
External Items, 68

F

Facebook, 257, 336
Filter, 210, 337
Finding matches, 183

Focus Group, 337
folders, 21
Fonts, 72
Framework method, 227, 337
funnel, 289
Fuzzy, 183

G

Google Chrome, 257, 262, 270
Graphs, 307
Grounded Theory, 337
Group Queries, 206

H

Help Documents, 331
hiding
  Annotations, 126
  Columns, 146, 203
  Picture Log, 112
  Rows, 145, 202
  See Also Links, 124
  sub-stripes, 293
  transcript rows, 103
  waveform, 98
Hushtag, 337

I

Import Project Report, 60
importing
  a Dataset, 241
  a Report Template, 325
  an Extract Template, 330
  Classification Sheets, 148
  Datasets, 241
  documents, 63
  media files, 92
  Picture-files, 107
  projects, 60
  Transcripts, 100
In Vivo Coding, 161, 337
inserting
  a symbol, 78
  a table, 77
  an image, 78
  date and time, 78
  page break, 78

Internet Explorer, 17, 257, 262, 270

J

Jaccard's Coefficient, 337

K

Kappa Coefficient, 291, 338
keyboard commands, 343

L

Last Run Query, 212
LinkedIn, 257, 338
Literature Reviews, 219

M

Map and Group, 255
Matrix Coding Queries, 196
Medline, 132, 338
Memo Link, 121
Memos, 117
merging
  Nodes, 133
  Projects, 60
  Transcript Rows, 100
MeSH terms, 132, 338
Mixed Methods, 338
Model Groups, 301

N

Navigation View, 48
NCapture, 257
Near, 183
New Attribute, 142
New Audio, 95
New Classification, 141
New Document, 66
New External, 68
New Folder, 22
New Framework Matrix, 229
New Hyperlink, 126
New Memo, 120
New Model, 297
New Node, 131, 132
New Relationship Type, 135
New See Also Link, 123
New Set, 30
New Video, 95
Nickname, 132, 154
node template, 294
NVivo Help, 331
NVivo Server, 295

O

OCR, 338
OneNote, 275, 338
Open Linked External File, 125
Open Referenced Source, 162
Open results, 211
Open To Item, 124
opening
  a cell, 201
  a Document, 66
  a Hyperlink, 127
  a Linked External Source, 124
  a Linked Memo, 122
  a Memo, 120
  a Node, 161
  a Picture Item, 110
  a Referenced Source, 162
  a See Also Link, 124
  a Video Item, 95
  an Audio Item, 95
  an External Item, 69
  an External Source, 70
Operators, 215

P

Page Setup, 82
Paste, 28, 204
Paste Special, 28
PDF documents, 85
Pearson Correlation Coefficient, 339
Phenomenology, 339
phrase search, 182
Picture Log, 110, 111
picture-files, 107
pie diagram, 172
Plain Text, 36
Preview Only, 211
Print List, 26

Print Options, 79
Print Preview, 79
Prohibit, 183
Properties
   Coding Comparison Query, 213
   Coding Query, 214
   Compound Query, 213
   Group Query, 206
   Matrix Coding Query, 213
   Model Group, 302
   Project, 51
   Shape/Connector, 302
   Subquery, 194
   Text Search Query, 213
PubMed, 132, 339

Q

Qualitative Research, 339
Quantitative Research, 339
Quick Access Toolbar, 31, 32
Quick Coding Bar, 153

R

Range Coding, 160
Read-Only, 167
Redo, 31
Refresh, 22, 23, 24, 26
RefWorks, 219, 339
Region, 110
Relationship Type, 134
Relationships, 134
Relevance, 183, 339
Required, 183
Research Design, 339
Reset Settings, 218
Ribbon, 32
Rich Text, 36
Root Term, 186
Run, 214

S

saving
   a Project, 62
   a Query, 210
   a Result, 211
Search and Replace, 74

Security Backup, 62
selecting text, 76
Service Pack, 333, 340
Sets, 29
Shadow Coding, 104
social media, 257
sorting
   Columns, 239
   Items, 285
   Options, 25
   Reports, 318
   Rows, 239, 285
Spell Checking, 74
Split Panes, 96
Spread Coding, 212
Static Model, 301
Status Bar, 19
Stemmed search, 182, 196
Stop Words, 340
Stop Words List, 179
subfolders, 21
Subquery, 194
Summary Links, 232
support, 332
Synonyms, 183
system requirements, 17
Sørensen Coefficient, 340

T

table, 77
Tag Cloud, 180
Team Members, 54
Teamwork, 287
Text Search Queries, 181
threshold value, 44
Tips for Teamwork, 294
Transcripts, 98
Tree Map, 180, 311
triple-click, 71, 76
Tutorials, 332
Twitter, 257, 340

U

Undo, 31
Undock, 27

unhiding
- *Annotations, 126*
- *Columns, 147, 203*
- *Picture Log, 112*
- *Rows, 145, 202*
- *See Also Links, 124*
- *sub-stripes, 293*
- *transcript rows, 103*
- *waveform, 98*

V,W

Validity, 340
Values, 139
video formats, 91
viewing
- *Coding Context, 165*
- *Coding Stripes, 167*
- *Excerpt, 164*
- *Highlighting Coding, 166*
- *Relationships, 137*

Wildcard, 183

Windows XP, 17
visualizations, 307
Wizard
- *Auto Code Dataset, 249, 267*
- *Chart, 169*
- *Classify Nodes from Dataset, 253*
- *Cluster Analysis, 308*
- *Extract, 326*
- *Import Classification Sheets, 149*
- *Import Dataset, 242*
- *Report, 319*
- *Tree Map, 311*

Word Frequency Queries, 178
Word templates, 158
Word Tree, 186

Z

Zoom, 78
Zotero, 219, 340

Made in the USA
Lexington, KY
13 April 2014